BIOMECHANICS – THEORY AND APPLICATIONS

BIOMECHANICS OF MARTIAL ARTS AND COMBATIVE SPORTS

BIOMECHANICS – THEORY AND APPLICATIONS

Additional books in this series can be found on Nova's website under the Series tab.

Additional E-books in this series can be found on Nova's website under the E-books tab.

SPORTS AND ATHLETICS PREPARATION, PERFORMANCE, AND PSYCHOLOGY

Additional books in this series can be found on Nova's website under the Series tab.

Additional E-books in this series can be found on Nova's website under the E-books tab.

BIOMECHANICS – THEORY AND APPLICATIONS

BIOMECHANICS OF MARTIAL ARTS AND COMBATIVE SPORTS

OSMAR PINTO NETO

Nova Science Publishers, Inc.
New York

LIBRARY OF CONGRESS CATALOGING-IN-PUBLICATION DATA

Neto, Osmar Pinto.
Biomechanics of martial arts and combative sports / Osmar Pinto Neto.
p. cm.
Includes index.
ISBN 978-1-61728-781-7 (softcover)
1. Martial arts--Training. 2. Self-defense. 3. Biomechanics. 4. Human locomotion. I. Title.
GV1102.7.T7N48 2010
796.8701'5--dc22
2010026619

Published by Nova Science Publishers, Inc. † New York

Dedicated to my family, especially Brisa, Carol, Iara, Iarinha and Osmar, Lee (my Kung Fu Master) and all my friends.

CONTENTS

PREFACE

By reading the title of this book, one might wonder what the difference between martial art and combative sport is. It is not easy to distinguish a martial art from a combative sport. For example, most people accept boxing to be a combative sport while Kung Fu to be a martial art. However, many Kung Fu schools participate in combative tournaments, and operate very similar to any standard boxing gym; they can be clearly considered combative sport schools. Trying to define what martial arts is may be helpful in clarifying possible differences between martial arts and combative sports.

The word "martial" from martial arts refers to the roman god Mars (the god of war); so martial arts could literally be translated as the art of war. In that sense, most styles that call themselves today martial arts are derived from ancient forms of combat, that were used at some point in time for war. If this concept encloses all the meaning of martial arts, then Boxing and Wrestling, for example, considered to most as combative sports and not martial arts, should also be considered as martial arts because the techniques from both these Olympic sports were also derived from ancient forms of combat used during wars. Clearly the strong sport competition orientation of both Boxing and Wrestling is not what differentiate them from traditional martial arts such as Judo (a Japanese martial art style) and Taekwondo (a Korean martial art style), which are also Olympic sports. So what do traditional martial arts have that set them apart from a combative sport? Honestly, from personal experience visiting and training in several martial arts and combative sports schools across Brazil, USA and China, I would not say that most traditional martial arts schools have one or more specific characteristics that are not present in Wrestling teams or Boxing gyms. To me, the main difference between what I understand as a martial art and a purely combative sport nowadays is that martial arts schools should teach not only combat techniques but also philosophical knowledge and principles that have been passed down from generation to generation. Martial arts have changed through time from "art of war" to "art of peace" and people that I consider martial artists are motivated by helping themselves and others to be better and happier human beings. Although I have seen coaches of combative sports that approach their sport exactly the same way I approach martial arts, it is not as common. So in the end, to me, there are martial artists that are solely combative sports athletes, and there combative sports athletes and/or coaches that are martial artists.

Nevertheless, being martial arts and combative sports the same thing or not, many people nowadays are practicing them, and many studies have been done trying to improve the performance of practitioners, understand different benefits for practitioners' health as well as understanding mechanisms of injuries and learning how to prevent them. Although there have been many epidemiological and health related studies of martial arts and combative sports, most of these studies are not reviewed in this book, unless they also involve biomechanics. As the title of the book suggests, the focus of this book is to review the advancements made in the biomechanics of martial arts and combative sports. The motivation of this book comes from a personal love for physics and martial arts since an early age. Biomechanics, in basic terms, is the area of science that makes the bridge between physics and human movements. Acknowledging that many martial artists and martial enthusiasts, who may be interested in this book, may not have a background in physics, I devoted the first chapter of the book to introduce biomechanics and expose some basic physical principles and biomechanical tools that are relevant to the studies reviewed in the book.

ABSTRACT

The pioneer studies on the biomechanics of martial arts were published in the 1960s and 1970s. After that, several other biomechanical studies have been conducted about martial arts and other related combat sports using a variety of different measures and methods, especially in the last decade. In general, these studies were concerned with: quantifying performance and investigating how to improve it; understanding of injury mechanisms and prevention; and investigating potential benefits from training martial arts and combative sports to the general population. This book presents a comprehensive review on this subject. It is divided in three major parts: the first part of the book covers the basics of biomechanics and clarifies how this field of science can help understand martial arts related questions; the second part of the book introduces a brief history of the most popular styles of martial arts and combative sports practiced around the world that have been previously studied in the field of biomechanics; the last part of the book presents a review of several articles including studies about the performance of specific hand strikes, kicks, throws and fall techniques, postural control benefits caused by martial arts training and biomechanical investigations of injury mechanisms and prevention.

INTRODUCTION

In the past 15 to 20 years the number of people training martial arts and combative sports has increased dramatically (Oler et al., 1991; Pieter and Lufting, 1994). Along with that, there has been also an increase in the number of scientific papers published about martial arts and combative sports. This book reviews the advancements made on the biomechanics of martial arts and combative sports.

Fights among unarmed men are probably as old as men themselves; in contrast, scientific studies of fighting are very recent. The pioneer studies on this subject were published in the 1960s and 1970s. Vos and Binkhorst (1966) were fundamental to set start the scientific investigations of martial arts with their article entitled "Velocity and force of some Karate arm-movements" published at Nature. During the 1960s, other important article was published, entitled "Karate strikes", by J.D. Walker (1975) in the American Journal of Physics. Finally, in 1978 an article entitled "The physics of Karate" by Feld, McNair and Wilk, published in the Scientific American, would contribute largely to the popularization of the studies on the biomechanics of martial arts.

After these articles were published, several other biomechanical studies have been conducted about martial arts and other related punching sports using a variety of different measures and methods, especially in the last decade. In general, these studies were concerned with: quantifying performance and investigating how to improve it; understanding of injury mechanisms and prevention; and investigating potential benefits from training martial arts and combative sports to the general population. This book is one of the pioneers on this subject. It is intended not only to biomechanics investigators looking for a review on what has and hasn't been already done, but also enthusiasts and practitioners of martial arts and combative sports. The first chapter of the book introduces the main biomechanical principles and methods of investigation used in the studies reviewed. The second chapter of the book is intended for readers curious about martial arts and combative sports history and exposes the most popular styles of hand-to-hand combat that have been previously studied. The following last three chapters of the book compile the review of over 100 studies conducted about the biomechanics of martial arts and combative sports.

Chapter 1

BASIC BIOMECHANICS

This chapter is intended for people with little previous knowledge of biomechanics. It exposes the history of biomechanics as well as its different areas.

Additionally, several physical definitions commonly used in biomechanical studies of martial arts and combative sports are explained.

1.1. KINESIOLOGY

Biomechanics is the field of science that investigates biological systems using mechanics. It is an extension of the field of science known as kinesiology. Kinesiology is the study of human movement (Knudson and Morrison, 2002).

Although the field of biomechanics has a somewhat short history (order of decades) kinesiology is centuries old, and it has been fascinating magnificence people throughout history.

Aristotle (384-322 BC) was one of them; his fascination with human movement was responsible for the first studies that described the action of muscles and classified those based on geometrical properties. Aristotle is considered to most the father of kinesiology (Enoka, 2008).

Aristotle's work served as base to several other great scientists, such as Galeno (129-199 AD), Galileu (1564-1642 AD), Borelli (1608-1679 AD) and, especially, Newton (1642-1727 AD).

1.2. Mechanics

Sir Isaac Newton was responsible for developing classical physics, or simply the mechanics. Newton's three laws of motion significantly changed the world. Drastic changes in pretty much all fields of science were caused by the applications of Newton's law. But it was not only science that was changed because of mechanics; society structure and the understanding of our universe were also molded by the laws of motion. Incredibly, for example, the industrial revolution and men's journey to the moon were both consequences of classical physics.

Mechanics can be divided in two different subareas: static and dynamic. Simply put, static mechanics studies the physics of bodies at rest. Dynamic mechanics, on the other hand, studies the physics of moving bodies. Dynamics can be subdivided in kinematics and kinetics. Kinematics is the study of the description of movements; it uses concepts such, as time, space, velocity, etc. Kinetics studies what causes movements, in such, it uses concepts such as mass, energy, power, etc (Hall, 1993).

Although mechanics has been known since the XVII century, it was only in the XX century around the 1960s that biomechanics was conceptualized as an area of science.

1.3. Physical Variables

In this section, a few physical variables that are commonly used in biomechanical studies of martial arts and combative sports are introduced.

1.3.1. Velocity and Speed

Velocity is a vector physical quantity (it has both magnitude and direction) that indicates the rate of change of position. In the International System of Units (SI), or metric system, it is measured in meters per second (m/s).

The velocity of a hand during a punch, for example, can be described as 5 m/s in the direction of the punching bag. In this example 5 m/s is the velocity's magnitude, whereas "in the direction of the punching bag" is the velocity's direction. The magnitude of velocity is normally referred as speed. There are two common types of velocities that are calculated in biomechanical studies: average and instantaneous velocities. Average velocity is defined as the ratio between a displacement (change in position) of a body (Δx) and the time it took for the displacement to happen (Δt). Instantaneous velocity is the velocity of a moving body for a particular instant of time. In other words, it is the velocity of a body when Δt is infinitesimally small.

1.3.2. Acceleration

Acceleration is the rate of change of velocity. Analogously, it is also a vector physical quantity. In the International System of Units (SI), or metric system, it is measured in meters per second squared (m/s2). The acceleration of a foot during a kick, for example, can be described as 2 m/s2 in the direction of the punching bag. There is no specific name for the magnitude of acceleration, as speed for example is used to represent the magnitude of velocity.

However, the term acceleration is commonly used to represent acceleration`s magnitude when there is an increase in speed, whereas the term deceleration is often used to represent a decrease in speed. Similarly to velocity we can define both average and instantaneous accelerations. Average acceleration is defined as the ratio between the change in velocity (Δv) and the time it took for this change to happen (Δt); whereas, instantaneous acceleration is the acceleration of a moving body for a particular instant of time.

1.3.3. Mass

In physics, mass (SI unit: kilogram - kg) commonly refers to any of three properties of matter, which have been shown experimentally to be equivalent: inertial mass, active gravitational mass and passive gravitational mass. In everyday usage, mass is often taken to mean weight, but in scientific use, they refer to different properties. Mass is an intrinsic property of matter, whereas weight is a force that results from the action of gravity on matter: it measures how strongly gravity pulls on that matter (weight equals mass times gravity (g); g is approximately equal to 9.8 m/s2 on Earth).

1.3.4. Linear Momentum

In classical mechanics, momentum (SI unit: kg.m/s or Newtons.s - N.s) is the product of the mass and velocity of an object. The principle of conservation of momentum states that the total momentum of any closed system (one not affected by external forces) is constant. Biomechanical studies of strikes that have used the principle of conservation of momentum to obtain results are common (Neto et al. 2007c; Walilko, 2005; Walker, 1975).

1.3.5. Force

Force (SI unit: N) is a concept that is used to describe the physical interaction of an object with its surroundings. It can be defined as an agent that produces or tends to produce a change in the state of motion of an object - that is, it accelerates the object (Enoka, 2008). For example, a punching bag hanging from the ceiling without moving (zero velocity) will remain at rest until a force act upon it.

In Biomechanics, the concept of strength (the ability of a person to exert force on physical objects using muscles) is important and known as the force which implies a certain human movement. Another kind of force is the impact force. This intense force acts during a certain minimum period of time in a collision.

When two bodies collide, the forces of interaction cause local deformations, heating and changes in motion of these bodies. During the collision, the magnitude of the impact force increases from zero to maximum during the period of contraction, and decreases from maximum to zero during the restitution period (McInnis and Webb, 1971).

Another very common force relevant to biomechanical studies of martial arts and combative sports is friction. Friction is the force resisting the relative lateral (tangential) motion of solid surfaces, fluid layers, or material elements in contact. The resultant horizontal component of the ground reaction forces between a shoe and the ground is an example of the friction force, also known as shear force.

The coefficient of friction is a dimensionless scalar value which describes the ratio of the force of friction between two bodies and the force pressing them together. The coefficient of friction depends on the materials used; for example, an old Wrestling shoe on a wet wood floor would exhibit low coefficient of friction, while a new Wrestling shoe on a rubber floor would exhibit high coefficient of friction.

1.3.6. Torque

The capability of a force to produce rotation is known as torque or moment of force. Torque (SI unit: N.m) is a vector that is equal to the magnitude of the force times the perpendicular distance between the line of action of the force and the axis of rotation (Enoka, 2008). All human movement involves the rotation of body segments about their joint axes; in other words, they can be described by torques.

1.3.7. Pressure

The distribution of force over an area is measured as pressure (P). The SI unit for pressure is the Pascal (Pa) which is equal to one Newton per square meter (N/m2). Non-SI measures such as pound per square inch (psi) and bar are commonly used in different parts of the world. A martial artist can increase the pressure of strikes by decreasing the contact area with the target. Conversely, a strike performed with gloves will produce less pressure for the same amount of force than a strike performed with no gloves.

1.3.8. Work (Mechanical Work)

In physics, Work (τ) is a scalar quantity that is calculated as the product of the displacement experienced by the object and the component of force acting in the direction of the displacement. The unit of work is joule (J; 1 J = 1 N.m). Work can be represented graphically as the area under a force-position (displacement) curve. This is a useful way to analyze movement when the force varies as a function of position (Enoka, 2008).

1.3.9. Kinetic Energy

Kinetic energy is a scalar quantity that depends on the mass of an object and its velocity squared. It is defined as the work needed to accelerate a body of a given mass from rest to its current velocity (SI unit: J). The kinetic energy of an object is the extra energy which it possesses due to its motion. Having gained this energy during its acceleration, the body maintains this kinetic energy unless its speed changes. Negative work of the same magnitude would be required to return the body to a state of rest from that velocity.

1.3.10. Power

Power (SI unit: watts - W) can be determined as the work done divided by the amount of time it took to perform the work, or as the product of average force and the speed. In short duration movements such as martial arts and combative sports strikes the rate at which muscles can produce work, or power production, is a critical performance variable.

1.3.11. Center of Mass and Center of Pressure

In studies that quantify stability as well as differences between regular gait and martial arts gait (Wu and Hitt 2005; Mao et al. 2006), two variables of key interest are the center of mass (also known as center of gravity) of the body and the center of pressure of the ground reaction forces. The center of mass of a system of particles or a body is the point at which the system's whole mass can be considered to be concentrated for the purpose of calculations. Forces applied through the center of mass of an unrestrained body generate zero torque, thus resulting only in translation but no rotation of the body (Benda, 1994). For example, a strike aimed near the center of mass of an opponent will be effective in terms of pushing your opponent back like a kick to the mid-section. On the other hand, a kick aimed far from the center of mass will be ideal to produce body rotation as in an ankle sweep, for example.

The center of pressure of a surface where forces of different magnitudes may act upon different areas of the surface is the average location of the pressure in the surface. We call the average location of the pressure variation the center of pressure in the same way that we called the average location of the weight of an object the center of gravity. In effect, the center of pressure is the location where the resultant force vector would act if it could be considered to have a single point of application (Benda, 1994). For the simple case where we assume a person is completely not moving, the vertical projection of the center of mass onto the floor coincides with the center of pressure. Most times however, this assumption is not true since the body sways even during stable standing. Center of pressure values have also been used in studies of martial arts and combative sports to determine the accuracy and precision of different strikes (Neto et al. 2009).

1.4. The Laws of Motion

Newton's laws of motion are three physical laws that form the basis for classical mechanics; one way or another most studies in biomechanics uses the laws of motion. These laws describe the relationship between the forces acting on a body and the motion of that body. They were first compiled by Sir Isaac Newton in his work Philosophiæ Naturalis Principia Mathematica, first published on July 5, 1687. Considering the understanding of the variables previously explained in this chapter, it is easier to understand Newton's three laws of motion. Although they have been expressed in several different ways over nearly three centuries, we can summarize the three laws as follows:

1st – In the absence of a net force (sum of all forces acting in a body, also known as resultant force), the center of mass of a body either is at rest (velocity equal to zero) or moves at a constant velocity (this law is commonly known as the "law of Inertia").

2nd – A body experiencing a net force F experiences an acceleration a related to F by $F = ma$, where m is the mass of the body (this law is often stated as, "Force equals mass times acceleration").

3rd – Whenever a first body exerts a force on a second body, the second body exerts a force on the first body of equal magnitude and opposite in direction (this law is commonly known as the "action-reaction law").

1.5. Main Sub-Areas of Biomechanics

In general, studies in biomechanics can be grouped into three main different categories: 1) studies related to the enhancement of performance in a sport; 2) studies related to reducing the injury risk of movements performed in sports and/or during normal day activities; and 3) studies that attempts to improve the mechanisms of rehabilitation. If we consider these three main categories of biomechanical studies and generalize three important concepts we can divide biomechanics into three main sub-areas: 1) performance; 2) injury; and 3) rehabilitation (Huston 2008). According to Huston (2008), performance refers to the way living systems do things. It connotes routine movements such as walking, sitting, standing, etc..., as well as global activities such as operating vehicles or tools, and sport mechanics. It also refers to internal movement and behavior such as blood flow, heart and muscle mechanics, and skeletal joint kinematics. Studies that aim the identification of the optimal technique for enhancing sports performance are very common and would fall into the performance category. Injury refers to failure and damage of biosystems as in broken bones, torn muscles, ligaments and tendons, and organ impairment (Huston 2008). The analysis of body loading to determine the safest method for performing a particular sport or exercise task as well as the analysis of sport and exercise equipment e.g., shoes, surfaces and racquets would fall into the injury category. Rehabilitation refers to the recovery of injury and disease; it includes all application of mechanics in health care industries (Huston 2008). Studies of muscular recruitment and loading during rehab are a common example of biomechanical studies that can be classified into the rehabilitation category.

There are several other ways to divide biomechanics into different areas. One approach is to consider the tools (variables) that are used to describe and analyze the movements of humans and other animals (Carpenter, 2005; Winter, 1990). Considering this approach we can divide biomechanics into 4 different main sub-areas: 1) Kinematics; 2) Kinetics; 3) Anthropometry; and 4) Electromyography.

1.5.1. Kinematics

As described before, kinematics is one of the areas of mechanics, which contains variables that are involved in the description of the movement, independent of forces that cause that movement. As such, it includes variables such as linear and angular displacements, velocities and accelerations. In biomechanics these variables are obtained from the displacement or acceleration data of anatomical landmarks such as the center of gravity of body segments, center of rotation of joints, extremes of limb segments, or key anatomical prominences (Winter, 1990). In turn, to obtain displacement data the most common tools used in biomechanics are motion tracking or motion capture systems and accelerometers.

Optical systems are the most common type of tracking system used in biomechanics. In simple terms, the general idea of optical system is that a subject wears markers that are followed by several cameras and the information is triangulated between them. Different systems can operate in different time resolutions. The time resolution of a motion capture system determines the number of samples (or photos) the system can obtain each second. When movements are performed in higher speeds, recordings need to be done in higher time resolutions in order to provide meaningful information. Some of the systems that can obtain information with high time resolutions are known as high-speed digital imaging systems or simply high-speed cameras. These systems are common in biomechanical studies of martial arts and combative sports because movements performed in martial arts and combative sports are faster than most other movements.

Accelerometers are devices that can measure the acceleration of a moving body relative to freefall. Conceptually, an accelerometer behaves as a damped mass on a spring. When the accelerometer experiences acceleration, the mass moves changing the conformation of the spring. The displacement of the mass is then measured to give the acceleration. There are many different types of accelerometers and each has unique characteristics, advantages and disadvantages. The different types include: piezo-electric accelerometers; piezo-resistive accelerometers; and strain gage based accelerometers. In many biomechanical studies of martial arts and combative sports accelerometers have also been used to estimate values of impact forces.

1.5.2. Kinetics

As described before Kinetics is also one of the main areas of mechanics, which contains variables associated to the causes of movement, such as force or torque. In biomechanics we can define, in relation to one particular subject, "internal forces" as forces that come from muscle activity, ligaments, or from friction in the muscle and joints. Additionally, we can define "external forces" to a subject as forces that come from the ground or from any external loads.

For example, when a boxer performs a punch, his/her muscles are responsible to produce force that will ultimately lead to an acceleration of his/her hand towards a possible target, like a punching bag.

The force produced by the boxer to accelerate his/her hand is an example of an internal force. When the boxer's hand collides with the punching bag, the punching bag applies a force onto the boxer's hand (please see next section under Newton's laws of motion). The force applied by the punching bag onto the boxer's hand is an example of an external force. From this example, it is clear that internal and external forces can be correlated. Since it is very hard to measure internal forces, most times in biomechanics internal forces are inferred from external forces that can be easily measured. The most common tools to obtain values of external forces in biomechanics are isokinetic dynamometers, load cells, pressure sensors and accelerometers (accelerometers are detailed in previous section).

An isokinetic dynamometer is a device designed for measuring specific human movement torques during speed controlled eccentric and concentric movements.

In view of its versatility and accuracy, the isokinetic dynamometer is suitable for both research and clinical use. However, because most martial arts and combative sports movements cannot be reproduced using an isokinetic dynamometer, studies about the biomechanics of martial arts and combative sports with dynamometers are scarce (Olsen et al. 2003).

A load cell, or force transducer, is an electronic device that is used to convert a force into an electrical signal. Although there are many varieties of load cells, strain gage based load cells are the most commonly used.

A strain gage load cell usually consists of four strain gauges that convert deformation (strain) into electrical signals. The electrical signal output of a typical load cell is in the order of a few millivolts and requires amplification by an amplifier before it can be interpreted.

In most martial arts and combative sports studies load cells are used to estimate impact forces. In order to that, however, in most cases, load cells need to be modified or adapted to other structures so they can accurately quantify the impact forces of punches or kicks without harming the strikers.

A pressure sensor is a device that, under constant force conditions, gives an output inversely proportional to the area of application of the force. There are several types of pressure sensors. Pressure sensors arranged as a plate, or board, are very common in biomechanical studies. These systems are important to quantify variables such as plantar pressure distribution and center of foot pressure, which are directly related among other things to body's balance.

Resistance-based force distribution sensors, such as those made by Tekscan Inc., have also been used in biomechanical studies of martial arts to quantify striking accuracy (Neto et al. 2009). They are appealing for biomechanics research because they are thin and flexible and offer high resolution and straightforward data acquisition. They also can provide real time dynamic feedback (Brimacombe et al. 2009).

1.5.3. Anthropometry

Anthropometry refers to the measurement of the human individual regarding masses and lengths of limb segments, locations of center of mass and center of rotation, angles of pull of muscles, moments of inertia, and so on. Many of the earlier anatomical studies involving body and limb measurements were not considered to be of interest to biomechanics. However, it is impossible to evolve a biomechanical model without anthropometric data. The accuracy of any biomechanical analyses depends as much on the quality and completeness of the anthropometric measures as on the kinematics and kinetics (Winter, 1990).

1.5.4. Electromyography

Electromyography (EMG) is a technique for evaluating and recording the electrical activity produced by skeletal muscles (Gary, 2004). When an end-plate potential is generated at a nerve-muscle synapse, it usually results in a muscle fiber action potential that propagates from the synapse to the ends of the muscle fiber. The currents associated with the muscle fiber potential can be measured with electrodes (Merletti and Parker, 2004). EMG is performed using an instrument called electromyograph, to produce a record called an electromyogram (also referred by the acronym EMG). EMG is used by clinicians to diagnose problems in the neuromuscular junctions, by ergonomists to determine the requirements of job-related tasks, by physiologists to identify the mechanisms involved in various adaptations within the neuromuscular system, and by biomechanists to estimate muscle force (Enoka, 2008).

Additionally, EMG is the most common technique used to quantify how the brain activates the muscles, being essential to investigate motor control and neuromuscular strategy under conditions. In other words, with EMG it is possible to obtain information regarding the final control signal of each muscle.

It gives information, for example, regarding which muscle or muscles are responsible for a muscle moment or whether antagonistic activity is taking place. Because of the relationship between a muscle's EMG and its tension, a number of biomechanical models have evolved. Among other things, EMG may also provide information regarding the recruitment of different types of muscle fibers and the fatigue state of the muscle (Winter, 1990).

Chapter 2

DIFFERENT STYLES OF MARTIAL ARTS AND COMBATIVE SPORTS

This chapter introduces a brief overview of the history of some of the most popular martial arts and combative sports that have been previously studied in the field of biomechanics. The biomechanical studies cited in this book are based on results coming from: Kung Fu, Tai-Chi Chuan, Karate, Taekwondo, Judo, Boxing, Wrestling and Mixed Martial Arts.

2.1. KUNG FU

Kung Fu is a popular term that expresses Chinese martial arts in general. Colloquially, the term kung fu in Chinese alludes to any individual accomplishment or cultivated skill obtained by long and hard work. The origins of Kung Fu can be traced over 6,000 years ago to self-defense needs, hunting activities and military training in ancient China (Chow and Spangler, 1982). From this beginning, Chinese martial arts proceeded to incorporate different philosophies and ideas into its practice expanding its purpose from simply military combat training to ultimately a method of self-cultivation. Today, there are hundreds of Kung Fu styles. Most of these styles were developed as progressions and offshoots of the styles created in two main ancient schools of martial arts in China (Reid and Croucher, 1983). These schools were located at two philosophical and religious centers, the Shaolin Temple (Buddhist) and the Wudang Mount (Taoist) (Despeux, 1981). Some styles focus on slower movements, meditation and breathing techniques and are labeled internal (e.g. Tai-Chi Chuan, Pa-Koua, Hsing-Hi Chuan), while others focus on faster movements, muscle and cardiovascular fitness and are labeled external (e.g. Wing Chun, Shaolin, Hung Gar). Another popular method of Kung Fu categorization is based on geography, as in northern and southern styles (e.g Northern Shaolin Praying Mantis and Southern Shaolin Praying Mantis). Northern styles are known to be fluid, elegant and dynamic. Southern styles on the other hand are known to be stable, powerful and grounded. These categories, however, are not perfect and many styles (e.g Yau-Man or Beggar Style) cannot be categorized as being neither internal/external nor northern/southern. Kung Fu started to spread internationally after the founding of the People's Republic of China in 1949. During that period, traditional Kung Fu styles also known as "family styles" were prohibited in China. Thus, many martial artists chose to escape from China and migrated to Taiwan, Hong Kong, United States, and other parts of the world. The wide world popularity of Kung Fu today can be attributed greatly to Hong Kong Kung Fu movies and a martial artist actor named Bruce Lee.

2.2. TAI-CHI CHUAN

Tai-Chi Chuan is the name given to several styles of Kung Fu that have very similar characteristics. The word tai-chi is an ancient Taoist philosophical term symbolizing the interaction of yin and yang, which are opposite manifestations of the same forces in nature. The dynamic interaction of yin and yang, underlying the relation and changing nature of all things, is depicted in the very well-known Tai-Chi Diagram. Tai-Chi is often translated as "grand extreme" and Chuan means fist or boxing. Thus, Tai-Chi Chuan is a pugilistic art rooted in the Taoist concepts of the interplay and necessary balance of yin and yang (Yang, 2005).

Tai-Chi diagram

The basic exercise of Tai-Chi Chuan is a series of defensive and offensive movements linked together in a continuous manner that flow smoothly from one movement to another (Perry, 1982; Smalheiser, 1984). These movements require continuous body and trunk rotation, flexion/extension of the hips and knees, postural alignment, and coordination of the upper and lower body (Swain, 1999). Deep breathing and mental concentration (or abstraction, as in meditation) are also required to evolve in Tai-Chi Chuan. Thus, Tai-Chi Chuan is not only a physical activity, but also one that involves the training of mental control (Kutner, 1997). Because of Tai-Chi Chuan characteristics and considering that mostly often it is trained at very slow speeds, Tai-Chi Chuan has been used for centuries in China as an exercise for health in a wide age range, particularly in the elderly.

The three major Tai-Chi Chuan styles are Yang, Wu and Chen. The Yang family style is the most popular and widely practiced style in the world today. Most biomechanical studies about Tai-Chi Chuan have been focused on investigating the potential benefits of Tai-Chi training to postural control and balance, especially for senior people.

2.3. KARATE

Karate is the name given to several Japanese styles of martial arts that involves primarily the use of punches, kicks and blocking techniques (Halabchi et al., 2007). Karate began as a common fighting system known as te among the samurai class of the Ryukyu Islands. Riukyu Islands are located west of China between Japan and Taiwan; Okinawa is its largest island. Still in its early history, two main factors furthered the development of Karate in Okinawa. The first, in 1372, was the establishment of trade relationships between Okinawa and the Ming Dynasty of China, which caused many forms of Chinese martial arts to be introduced to Okinawa. The second was the "Policy of Banning Weapons", enforced in Okinawa during the late 1400's. (Reid and Croucher, 1983). Because of the Chinese influence, the word karate was originally a way of expressing "Chinese hand". In 1933, the Okinawan art of Karate was recognized as a Japanese martial art and Karate started to be written in Japanese characters now that translate as "empty hand" (Higaonna, 1985).

There are four major styles of Karate today: Shotokan-ryu, Wado-ryu, Goju-ryu and Shito-ryu (Thompson, 2008). Shotokan-ryu was one of the first styles to be introduced to Japan in the 1920's. It is best characterized by its long and deep stances and its use of more linear movements. Wado-ryu is a derivative of Shotokan-ryu karate, Jujutsu, grappling and Tai Sabaki (Body Movement). Wado-ryu karate does not practice many of the body toughening exercises common to other styles of karate, preferring rather to use Tai Sabaki to evade attacks. Goju-ryu is known by utilizing up and down stances and internal breathing power (known as "hard and soft" techniques). In simple terms, it is a combination of Shotokan-ryu and Chinese White Crane Kung Fu. Shito Ryu was formed by the combination of the kata and techniques of two older styles of Karate, Naha-Te (which developed into Gojuryu Karate) and Shuri-Te (which developed into Shorinryu). Characteristic for Shito-Ryu Karate are the square-on stances, linear strikes and handling of traditional Okinawan weapons. Karate is one of the most widespread martial arts in the world. Due to its early wide popularity in the United States, Karate was the first martial art to be investigated scientifically (Vos and Binkhorst, 1966).

2.4. Taekwondo

Taekwondo is a Korean martial art very popular worldwide and the national sport of South Korea. In Korean, tae means foot; kwon means fist; and do means way. Traditional Taekwondo is typically not competition-oriented and stems from military roots. Modern Taekwondo, which is much more popular, on the other hand, tends to emphasize competition. Formally, there are two main styles of Taekwondo. Although there are doctrinal and technical differences between the two main styles and among the various organizations, the art in general emphasizes kicks. The roots of Taekwondo go back thousands of years; however, only in 1955 Taekwondo was officially established when several similar Asian schools of martial arts were merged. An important figure in this effort was Choi Hong Hi, a Korean general who worked to combine a traditional Korean foot-fighting technique called tae kyon with Japanese Karate. In 1966, General Choi established the International Taekwondo Federation (ITF) (Taekwondo, 2008).

2.5. Judo

Judo is a modern Japanese martial art and combat sport that originated in Japan in the late nineteenth century. Judo was developed by Jigoro Kano

(1860–1938). In 1880, Kano reformed an ancient Japanese martial art called Ju-Jutsu with a new set of rules for competition and with focus on development of the body, mind and character of the practitioners. The word judo is formed by two characters: ju, which means soft method, and do, which means, as in Taekwondo, way, road or path (Group, 1977). Unlike Kung Fu, Karate and Taekwondo, there is only one main style of Judo. Its most prominent feature is its competitive element, where the ultimate goal is to either throw one's opponent to the ground at his back, or immobilize or force one's opponent to quit with grappling maneuvers (Takahashi, 1982). Most biomechanical studies about Judo have been focused on investigating the possibilities of teaching Judo fall techniques to non-athletes adults, especially senior people, in an attempt to reduce hip fracture risk from fall.

2.6. Boxing

Boxing, also known as Western Boxing or pugilism, is a combat sport in which two participants, generally of similar weight, fight each other with their fists wearing padded gloves. Boxing first appeared as a formal Olympic event in the 23rd Olympiad (688 BC), but fist-fighting contests must certainly have had their origin in mankind's prehistory. The earliest visual evidence for Boxing appears in Sumerian relief carvings from the 3rd millennium BC (boxing, 2008). Modern Boxing evolved in Europe, particularly Great Britain. During the early 18th century, in Great Britain, Boxing was done bare-knuckle and referred to as prizefighting. The first documented account of a bare-knuckle fight in England appeared in 1681 in the London Protestant Mercury. This is also the time when the word boxing first came to be used (Roberts and Skutt, 2002). Pugilism is a much older word, it indicates the ancient origins of the sport in its derivation from the Latin pugil, which mean a boxer, related to the Latin pugnus, which means fist, and derived in turn from the Greek pyx, which means with clenched fist (boxing, 2008).

Boxing is supervised by a referee and is typically engaged in during a series of one to three-minutes rounds. Victory in Boxing can be achieved in three different ways: by points at the end of the bout, given by judges' scorecards; by knockout, when a fighter is knocked down and unable to get up before the referee counts to ten seconds; or by technical knockout, when a fighter is considered by the referee not to be able to continue (Group, 1977).

2.7. WRESTLING

The term wrestling is an old English word that originated some time before 1100 AD. It is perhaps the oldest word still in use in the English language to describe hand-to-hand combat.

Nowadays, Wrestling is a combative sport where there is a physical engagement between two unarmed persons, in which each wrestler strives to get an advantage over or control of their opponent. Physical techniques which embody the style of Wrestling are clinching, holding, locking, and leverage. No striking is permitted in modern Wrestling.

The modern sportive form of Wrestling derives from prehistory survival fighting (Arlott, 1976). There is considerable evidence that Wrestling existed in all early civilizations.

In parts of southern Europe, evidences of Wrestling's origins has been found in paintings and carving in caves. Many of these paintings and carves date from 20,000 years ago. Artifacts have also shown that Wrestling has been around in several ancient civilizations, like Mesopotamia, Khafaji (near present-day Baghdad, Iraq), Egypt and Greece (Chiu, 2005).

Nevertheless, it was in ancient Greece that it really developed into a sport, and was included in the Ancient Olympic Games in 704 BC (Arlott, 1976).

According to the International Federation of Associated Wrestling Styles (FILA), there are several International Wrestling disciplines acknowledged throughout the world, including the Greco-Roman Wrestling, the Freestyle Wrestling, and the traditional or folk styles of Wrestling.

The Greco-Roman style of Wrestling was developed in France and does not allow holds on the legs or any form of tripping. It became an event at the first modern Olympic Games, in Athens in 1896. Freestyle Wrestling developed in Great-Britain and in the United States under the name of catch-as-catch-can. Freestyle Wrestling became an Olympic event in 1904.

Women's freestyle Wrestling was added to the Olympics in 2004. There are several different folk-styles of Wrestling. Today USA Collegiate Wrestling is the most popular of them. Collegiate Wrestling emerged out of the folk Wrestling styles practiced in the early history of the United States.

The ultimate goal in Collegiate Wrestling as in freestyle Wrestling is to pin the opponent to the mat, which results in an immediate win. Collegiate and freestyle Wrestling, unlike Greco-Roman Wrestling, allow the use of the wrestler's or his/her opponent's legs in offense and defense.

2.8. Mixed Martial Arts

Mixed Martial Arts (MMA) is a combative sport where two unarmed persons can use techniques from several other martial arts and combative sports to try to either knock out or submit an opponent (make the opponent quit). Although, the term MMA, and the sport as we know today are relative new, the idea of mixing more than one martial art to create a new one, is probably as old as martial arts styles themselves. One great example of the mixing of two very different martial arts into a single combative sport very similar to modern MMA happened in Greece in 648 BC, when the Greeks introduced the sport of Pankration into the Olympic Games. The word Pankration is a combination of two Greek words, pan, meaning all, and kratos, meaning powers. This is an accurate depiction of the sport itself, as it was a mixture of ancient Greek boxing and Wrestling. The sport became the most popular event in the Olympic Games. However, with the rise of the Roman Empire there was a decline of Pankration in Greece (Walter, 2003).

The beginning of modern MMA can be traced back to a family named Gracie in the early 1900s Brazil. The Gracie brothers (mainly Carlos and Helio Gracie) were responsible for adapting Judo/Ju-Jutsu, which Carlos had learned from a Japanese master named Mitsuyo Maeda, or Count Koma, and creating what is today known as Gracie Jiu-Jitsu or simply Brazilian Jiu-Jitsu (Gracie, 2008). To promote their style, the Gracie brothers created the "Gracie Challenge", in which they challenged fighters of other styles of martial arts to fight with the minimum amount of rules. In Brazil these fights were known as Vale-Tudo, which in Portuguese implies that the fights had no rules. These challenges became popular world-wide when Brazilian Jiu-Jitsu fighter Royce Gracie won the first Ultimate Fighting Championship tournament in the United States in 1993. Royce Gracie won the tournament by submitting all his opponents using grappling maneuvers without hurting either himself or his opponents during the fights. Gracie Jiu-Jitsu fighters demonstrated a clear advantage over other opponents with experience in only one martial art style and no previous Jiu-Jitsu experience; thus, they motivated other fighters such as wrestlers and striking artists (e.g. boxers and Muay-Thai specialists, Muay-Thai is a traditional Thailand's kickboxing fighting style, which among other things are well-known for strikes with knees and elbows) to start cross-training (training more than one style). Today most MMA fighters train a mixture of Boxing, Muay-thai, Wrestling and Brazilian Jiu-Jitsu.

Chapter 3

Kinetics and Kinematics Qualification of Martial Arts and Combative Sports

Since the beginning of martial arts biomechanics research, scientists were intrigued in quantifying the speeds and forces involved in strikes performed by highly trained subjects. Recently a few of studies have compared different strikes to determine which type of strike would be the best for different fighting situations. A few studies have also been conducted trying to unveil the differences between trained and untrained subjects' performances.

3.1. Movement Velocity and Impact Forces

3.1.1. Hand Strikes

Vos and Binkhorst (1966) were one of the first scientists to investigate the physics of martial arts strikes. Their main interest was better understanding how Karate practitioners are able to break bricks, tiles or blocks during demonstrations. They collected kinematic and kinetic data from 3 highly experienced Karate practitioners and 2 control subjects with no karate experience. Participants attempted to break bricks and plates using their bare hands. Karate experts were able to achieve higher hand speeds than control subjects. The maximum hand speed reported was 51 km/h or 14.2 m/s. The impact forces varied depending if the objects broke or not during the collision. Both Karate experts and control subject succeeded and failed in different breaking attempts. The authors overall conclusion was that Karate experts develop higher hand speeds and courage than non-trained subjects causing them to be more efficient in breaking demonstrations.

Joch et al. (1981) was one of the first studies to quantify the impact force of punches targeting non-breakable objects. They measured average punch impact forces of 3453 N for elite and 2932 N for intermediate boxers. Wilk et al. (1983) also interested investigated Karate breaking demonstrations. They presented a kinematical study and a dynamical theory of Karate techniques for patio blocks breaking. In their study, strobe photographs were taken at 60 or 120 flashes/s and high-speed videos (1000 - 5000 Hz) were obtained to analyze Karate movements and hand collision with patio blocks. They reported that adept Karate practitioners could obtain maximum speeds values in the range of 5.7-9.8 m/s when performing straight punches and 10-14 m/s when performing downward hammer-fist or knife-hand strikes. They also reported impact forces ranging from 2400-2800 N. Atha (1985) gathered data from a world ranked professional heavyweight as he punched an instrumented, padded target mass suspended as a ballistic pendulum. Within 0.1 s of the start the punch had travelled 0.49 m and attained a velocity on impact of 8.9 m/s. The peak force on impact was 4096N, attained within 14 ms of contact. The transmitted impulse generated an acceleration of 520 m/s2 (53 g) in the target. Voigt (1989) used a custom-built dynamometer (accelerometers) as well as stroboscopic photography (100 Hz) and high-speed video (500 Hz) to analyze punches of 10 well trained Karate students. He found peak impact force of 3334 N (2345 - 4866 N), and maximal speed of the hand before impact of 9.5 m/s (8.2 - 10.7 m/s). Smith et al. (2000) measured average impact forces, using a sport-specific designed dynamometer, of 4800 N (SD = 601) for elite and 3722 N (SD = 375) for intermediate boxers.

3.1.2. Kicks

There are not as many studies on the biomechanics of kicks as there is of hand strikes. Wilk et al. (1983) measured the speed of 5 different Karate kicks and obtained values ranging from 7.3-14.4 m/s. Pieter and Pieter (1995) measured forces up to 620 N for the round kick against a water-filled bag with a built-in force sensor unit. Conkel, et al. (1988) used piezoelectric film, similar to that used to measure pressure distribution on the foot during gait, and attached it to a heavy bag to measure the impact force of front, side, back and roundhouse kicks. Impact forces of up to 470 N were recorded for the roundhouse kicks. Pieter and Pieter (1995), Conkel, et al. (1988) and Serina and Lieu (1991) have measured linear velocity of the foot immediately before impact during the execution of the round kick ranging from 14 to 16 m/s. Boey et al. (2002) reported kinematical results of turning kick performance of 4 athletes, two men and two women, in the Singapore National Taekwondo squad. Kicks speeds ranged from 11.53 to 15.04 m/s for the women and 17.12 to 22.7 m/s for the men. Pedzich et al. (2006) investigated the side-kick and spinning back kick performed by 5 competitors with mastery class of Taekwondo.

They reported average maximal stroke force values of 9015±2382 N (right leg) and 8294±2308 N (left leg) for the side-kick and 8569±2381 N (right leg) and 7751±2570 N (left leg) for the spinning back kick. Tsai et al. (2007) reported values of ankle speed of 9.55±1.65 m/s for the spin-whip kick performed by eight elite Taekwondo male college students.

3.1.3. Throws

To date, only a handful of studies have investigated Judo from a biomechanical perspective. Harter and Bates (1985) studied the ground reaction forces associated with the harai-goshi throw. A tri-modal peak anteroposterior ground reaction forces pattern indicating a pull-push-pull effort by thrower was found during the throw. Tezuka et al. (1983) also measured the ground reaction forces of hara-goshi and found similar results.

Imamura et al. (2007) was the first article to consider the biomechanics of Judo throwing out of isolated laboratory conditions. They compared the kinematics of throwing under competitive and non-competitive conditions under a real-life competitive condition and a simulated laboratory condition.

Their results were based on data collected from a third degree black belt subject served as the thrower for both conditions, and two black belt participants ranked as first degree and fourth degree that served as the faller.

They found that, in general, the directional velocity patterns of the peak velocity for the center of mass of the thrower and faller were similar in both conditions.

During competition, thrower created larger peak center of mass directional velocities onto the faller, which indicated greater throwing power. Peak velocities for thrower center of mass were larger during the noncompetitive condition since faller resistance was minimal.

3.2. Direct Comparison between Different Techniques

3.2.1. Hand Strikes

Gulledge and Dapena (2008) compared mechanical factors in the reverse and three-inch power punches performed by twelve expert male Karate martial artists. They used a force plate to measure horizontal peak forces, and subsequently impulses. The power punch produced smaller fist velocities immediately before impact than the reverse punch, 4.09 vs. 6.43 m/s. The peak force exerted by the fist was much smaller in the power punch than in the reverse punch 790 vs. 1450 N. However, they found that the linear impulse exerted by the fist during the first 0.20 s of contact was slightly larger in the power punch than in the reverse punch (43.2 vs. 37.7 N.s). They concluded that the power punch is less powerful than the reverse punch, but slightly more effective for throwing the opponent off balance. Bolander et al. (2009) investigated different applications of strikes from Kung Fu practitioners that had not been addressed before in the literature. Punch and palm strikes were directly compared from different heights and distances, with the use of a load cell, accelerometers, and high-speed video. The data indicated that the arm accelerations of both strikes were similar, although the force and resulting acceleration of the target were significantly greater for the palm strikes. Additionally, the relative height at which the strike was delivered was also investigated. The overall conclusion was that the palm strike was more effective than the punch for transferring force to an object and that an attack to the chest would be ideal for maximizing impact force and moving an opponent off balance.

3.2.2. Kicks

O'Sullivan et al. (2009) also examined the effect of different heights on striking impact force. In this study however the authors compared Taekwondo and Yongmudo (YMD) turning kick. Yongmudo is a martial art that has been developed by Yong-In University professors and it was created by combining and modifying the techniques of several other martial arts styles. They analyzed kicks performed by 5 skilled practitioner of each style. To measure the impact force, they fixed two accelerometers to a PVC pipe in a sandbag.

Each participant performed 10 turning kicks trunk and face height in random order. As did Bolander et al. (2009) have found for hand-strikes, O'Sullivan et al. (2009) found that kicks to the mid section were stronger than those to the high section.

3.2.3. Throws

Imamura et al. (2006) recorded and analyzed four black belt throwers using two video cameras (JVC 60 Hz) and motion analysis software. They investigated three different throwing techniques: harai-goshi (hip throw), seoi-nage (hand throw), and osoto-gari (leg throw). Each throw was broken down into three main phases; balance breaking, fit-in and actual throw. Their results showed that for the harai-goshi and osoto-gari throws, impulse measurements were the largest within the first two phases of the movement. The seoi-nage throw demonstrated the lowest impulse and maintained forward momentum on the body of faller throughout the entire throw.

3.3. Differences in Performance between Trained and Untrained Subjects

3.3.1. Hand Strikes

Several articles have made contributions in the attempt to understand how martial artists could obtain such values of force and power as described above, and why and how they perform different than untrained subjects; most of them concerns hand strikes. Wilk et al. (1983) suggested that the hand speed right before the impact was the primary factor contributing for the greater impact force of a Karate martial artist strike compared to a non-martial artist. However, they did not verify this suggestion in their study. Walker (1975) suggested that another important variable, the effective mass of impact, could vary in different forms of Karate strikes and affect the impact force; however, he did not report values of this variable.

The effective mass of impact is a measure of a body's inertial contribution to the transfer of momentum during a collision. In the case of a martial art strike, the effective mass can be seen as the mass of an imaginary rigid body that could replace the striker and with the same speed as the hand speed before the impact produce the same effect on the collision as the striker would. Blum (1977) suggested that an adept Karate practitioner achieves a "high mass" by tightening all the appropriate arm and upper body muscles at the moment of impact, but further insight into this theory was not provided (please see Chapter 5: *section 5.1.2*).

Smith and Hamill (1986) were the first authors to investigate if higher skilled martial artists can exhibit higher values of effective mass than lesser skilled martial artists. They measured the fist velocities from Karate athletes of different skill levels and the relative momentum of a 33 Kg punching bag. The bag momentum was greatest for the highest skilled subjects compared to the lower skilled punchers even though their respective fist velocities were approximately the same (11.03 (standard deviation (SD) 1.96) m/s for all subjects). Smith and Hamill (1986) suggested that the increase in bag momentum was due to the skilled boxer's ability to generate a greater effective mass during the impact than the lower skilled boxers. The estimated average effective mass for the highest skilled boxers was approximately 4.1 Kg. Since this value is greater than the mass of the hand, the authors believed it reflected the ability of the athletes to link the mass of the arm into the punch. Voigt (1989) found the effective mass of Karate punches calculated from calibrated punching dynamometer to be 1.4 Kg (1.0 - 2.2 Kg) and calculated from stroboscopic recordings to be 4.0 Kg (2.5 - 4.5 Kg). Walilko et al. (2005) reported a mean effective punch mass of 2.9 (SD 2.0) Kg and a slight linear association of the effective mass with the weight of the boxer.

Neto et al. (2007c) was the first study that calculated values of effective mass from both trained and untrained subjects. In their study, the subjects were asked to strike a basketball at rest on a table, five times, with maximum force using a palm strike. Kung Fu athletes presented significantly higher averages of effective mass, hand speed and performance (ball speed after collision) than control group. The average effective mass for the Kung Fu group was 2.62 (SD 0.33) Kg and for the control group was 1.33 (SD 0.19) Kg. The average hand speed was 6.67 (SD 1.42) m/s for the Kung Fu group and 5.04 (SD 0.57) m/s for the control group. The average ball speed was 9.00 (1.89) m/s and 5.72 (SD 0.44) m/s for the Kung Fu group and control group, respectively. Comparing the values obtained for the average effective mass and hand and forearm mass estimated using regression equations, the authors found that for the control group the effective mass was statistically equivalent to the mass of the hand and forearm, while for the Kung Fu group the effective mass was significant greater than the mass of the hand and forearm. Contrary to what was found by Walilko et al. (2005), for both groups and all subjects no significant correlation was found between the values of effective mass and body mass. The authors concluded that because of the higher effective masses of the KF athletes, the KF group performance increases approximately twice as much with the increase of hand speed than the performance of the control group and that the effective mass is the main factor that distinguish a trained from a non-trained subject. Accordingly, Neto et al. (2008b) found a significant linear relationship between the values of average muscle and impact power for experienced Kung Fu Yau-Man practitioners, but not for the non-experienced participants. In other words, for the experienced participants an increase in muscle power was directly reflected in an increase in impact power, while for the non-experienced participants a higher muscle power did not necessarily corresponded to a higher impact power.

3.3.2. Kicks

The biomechanical comparison between novice and experienced martial artists performing kicks are still very rare. Falco et al. (2009) investigated one of the most frequently used kicks in Taekwondo competition, the Bandal Chagui or roundhouse kick. Because excellence in Taekwondo relies on the ability to make contact with the opponent's trunk or face with enough force in as little time as possible, while at the same time avoiding being hit distance between contestants is an important variable to be taken into consideration. Thus, Falco et al. (2009) examined both impact force and execution time in the roundhouse kick performed by 31 Taekwondo athletes (both expert and novice, according to experience in competition) and explored the effect of execution distance in these two variables. A force platform and a contact platform were used to measure these variables. The study showed higher maximum impact forces for expert competitors than for novice competitors. Also, expert competitors were more powerful in longer distances than novice competitors in the closest. The results showed that there are no significant differences in terms of impact force in relation to execution distance for the expert competitors, but the differences were significant for the novice competitors. They also found significant and positive correlations between body mass and impact force only for the novice athletes. This result corroborates with the effective mass results found for the hand strikes reviewed above suggesting that martial arts experience can alter effective mass and demonstrates that for the experienced athletes technique played a larger role than actual mass in their performance.

3.3.3. Throws

Pucsok et al., (2001) analyzed and compared the kinetic and kinematic characteristics of the throwing technique harai-goshi (hip throw), of novice and advanced Judo competitors. Kinetic and kinematic data were collected by utilizing the Kistler Instrument Corporation Multicomponent Force Measuring Platform System and the Peak Technologies Motion Video Analysis System. This article revealed a significant difference in horizontal force application, between novice and advanced Judo competitors. Numerous significant relationships among mean horizontal ground reaction force application and horizontal leg sweep velocity in was found in 19 of the 28 participants of the study. They suggested that leg sweep velocity is a function of ground reaction force application and horizontal leg sweep velocity plays a primary role in good technical execution of the harai-goshi throw.

Imamura and Johnson (2003) recorded twenty male Judo players, 10 black belts and 10 novices, executing the major outer leg reap with maximal effort. Statistical analysis found that peak angular velocity of the trunk and peak angular velocity of the ankle differed significantly from the performance of the experienced and novice participants. These differences were attributed to good upper body to upper body contact or impact, and stressed the importance of executing plantar flexion near sweep contact. Their results emphasized the importance of using the sweeping leg in a sequential kinetic link motion rather than as a single rigid segment.

Chapter 4

INJURY ACESSMENT AND PREVENTION

Like in most other sports, injuries are very common in martial arts and combative sports. Although not reviewed in this book, there have been several non-biomechanical studies that reported injury trends, injury case reports, pathologic anatomy of specific injuries, as well as injury prevention strategies for different martial arts and combative sports (e.g. Baker et al. 2010; Melone et al. 2009; Ngai et al. 2008). The focus on this chapter, however, is to review biomechanical studies related to injuries common in martial arts and combative sports. Such studies can be divided in two main groups: 1) studies that attempt to quantify important variables that can help understand the causes and extent of injuries; 2) and studies related to the prevention of injuries.

4.1. INJURY ASSESSMENT

4.1.1. Strikes

The values of force of impact from hand strikes shown in the previous chapter were not estimated considering impacts to objects comparable in mass and bio-fidelity to segments of the human body, thus the risk of injury cannot be reliably estimated from those forces. In an effort to better understand the relationship between forces delivered to human body and the risk of injury, a few studies have been conducted. Schwartz et al (1986) investigated the relative force of kicks and punches thrown at a dummy head, which was mounted 175 cm above the floor, their mechanism was contrived to provide constant rotational stiffness, and springs provided constant restorative moments about the three axes. They asked fourteen Karate experts to punch and kick the dummy and measured accelerations ranging from 90 to 120 g. Walilko et al. (2005) submitted Olympic class boxers to perform straight punches thrown at an instrumented Hybrid III headform that simulated a human head. They calculated impact force and impact power values for seven boxers' punches and obtained an average impact force of 3427 N (SD = 811) and an average impact power of 6574 W (SD = 3453); they reported an average hand speed of 9.14 m/s (SD = 2.06). Pierce et al., (2006) obtained the first direct measurement of punch force in professional Boxing matches through Boxing gloves incorporating the bestshot SystemTM, a proprietary system created by SensorPad Systems, Inc. (SPS; Norristown, PA). They obtained force data across all rounds of six professional Boxing matches across five different weight classes. Mean punch forces delivered ranged from 866.6 N to 1149.2 N. These measurements are considerably less than those reported in most other research performed in laboratory settings, and according to Pierce et al., (2006) may better reflect the actual forces exchanged in the ring. More studies should be done in order to verify if these forces do represent better the forces exchanged during actual Boxing matches.

Stojsih et al. (2008) also obtained biomechanical measures during live boxing. Several biomechanical factors of the head during a sparring session and their link to cognitive function were evaluated. Data was obtained by "The Impact Boxing Headgear" (IBH), a system consisting of a head gear equipped with 12 single axis linear accelerometers, a wireless transceiver, onboard memory, and data acquisition capabilities (1,000 Hz). Data were collected from 30 male and 30 female amateur boxers during 4-two minute sparring sessions. Neurocognitive assessment was measured using the ImPACT© Concussion management software. A baseline neurocognitive test was obtained from each athlete prior to sparring; two additional tests were obtained and compared to the baseline. They found that Peak translational and rotational acceleration values of the head were 191 g and 17,156 rad/s2, respectively, for males and 184 g and 13,113 rad/s2, respectively, for females. The peak Head Injury Criteria (HIC) and Gadd Severity Index (GSI) values for males were 1,652 and 2,292, respectively, and for females were 1,079 and 1,487, respectively. There was no significant difference in the neurocognitive scores between genders. In conclusion, the majority of impacts experienced by both genders were under the threshold for mild head injury. However, we should consider that the impacts measured in this study happened during training sessions, and that strikes may be stronger during actual tournament fights.

4.1.2. Takedowns

Injuries during martial arts and combative sports however are not only caused by strikes. Takedowns, which are maneuvers performed by a fighter to put the opponent on the ground, can also cause many injuries. Several different types of takedown are normally performed during Mixed Martial Arts fights. MMA is the sport that grows faster in the world and has rapidly succeeded Boxing as the world's most popular contact sport (Kochhar et al. 2005). The incidence of injury in this combative sport is recognizably high (Bledsoe et al. 2006; Ngai et al. 2008). However, only one biomechanical study that access possible injuries during MMA was found. Kochhar et al. (2005) proposed a study to assess qualitatively and quantitatively the potential risk for participants to sustain cervical spine and associated soft tissue injuries due to MMA takedowns. They analyzed four commonly performed takedowns (the hip toss; the suplex; the souplesse; and the guillotine drop) with possible risks to the cervical spine with respect to their kinematics, and constructed biomechanical models. Their kinematics results showed that all takedowns presented a strong risk of injury. They also showed similarities in the force, kinematics, and biomechanics required to produce cervical neck injuries in rear impact vehicle accidents and these four common martial arts maneuvers. Their mathematical models showed that the forces involved are of the same order as those involved in whiplash injuries and of the same magnitude as compression injuries of the cervical spine.

4.2. Injury Prevention

The biomechanical studies of martial arts related to injury prevention can be divided in two different categories. The first category include studies that investigate the benefits of training martial arts in preventing injuries; this category can be divided in studies that have investigated how martial arts training can help improving balance and postural control and studies that that have investigated how martial arts techniques can help reducing the forces exerted in the human body during falls. The second category includes a study that investigates how biomechanical characteristics of the shoes and mats used by wrestlers during competition may increase or decrease their risk of injury.

4.2.1. Balance and Postural Control

Studies have indicated that Tai-Chi Chuan increases postural stability and enhances balance (Hong et al., 2000; Thornton et al., 2004; Tsang et al, 2004a,b), which leads to a reduction in the risk of falls (Li et al. 2004).

In a cross sectional study by Hong et al. (2000), 28 male Tai-Chi Chuan practitioners with an average age of 67.5 ± 5.8 years old and 13.2 years of Tai-Chi Chuan exercise experience were compared to 30 sedentary men aged 66.2 ± 6.5. Measurements included, among some general fitness tests, left and right single leg stance with eyes closed. Compared with the sedentary group, the Tai-Chi Chuan practitioners demonstrated to have better balance.

Tsang et al. (2004a) further investigated the effects of long-term Tai-Chi practice on balance control when healthy elderly Tai-Chi practitioners stood under reduced or conflicting somato-sensory, visual, and vestibular conditions. They compared twenty elderly experienced Tai-Chi practitioners (7.2 ± 7.2 years of experience) with 20 healthy elderly non–Tai-Chi practitioners and 20 healthy young subjects. They used computerized dynamic posturography to measure the amplitude of antero-posterior body sway under different somato-sensory, visual, and vestibular conditions. The results demonstrated that the Tai-Chi practitioners had significantly better balance control than the non–Tai-Chi subjects. Even more interesting, they found that the elderly Tai-Chi Chuan practitioner achieved a level of balance performance comparable to that of young healthy adults.

Different than in the previous studies where long term effects of Tai-Chi Chuan were investigated, Tsang et al. (2004b) examined whether 4 and/or 8 wk of intensive Tai-Chi practice would be enough time to improve balance control in the healthy elderly subjects. Forty-nine community dwelling elderly subjects (aged 69.1 ± SD 5.8 years old) voluntarily participated in their study participating in an intervention program (1.5 h, 6 times a week for eight weeks) of either supervised Tai-Chi or general education. Two balance tests were administered using computerized dynamic posturography before the intervention, and after 4 and 8 weeks of training. They tested subjects' abilities to use somato-sensory, visual, and vestibular information to control their body sway during stance under six sensory conditions and measured the subjects' abilities to voluntarily weight shift to eight spatial positions within their base of support. They found that after 4 and 8 weeks of intensive Tai-Chi training, the elderly subjects achieved significantly better vestibular ratio in the sensory organization test and improved directional control of their leaning trajectory in the limits of stability test, when compared with those of the control group. Furthermore, they showed that the improved balance performance from week 4 on was comparable to that of experienced Tai-Chi practitioners, indicating that 4 weeks of intensive Tai-Chi training are sufficient to improve balance control in the elderly subjects.

There are not many studies that have attempt to understand the mechanisms of Tai-Chi Chuan by addressing why it may be a better exercise for improving balance and posture control than other traditional balance exercises (Hong and Li 2007).

Xu et al. (2003) analyzed the kinematics and muscle activity of a typical Tai-Chi Chuan movement (brush knees and twist steps) with the purpose of investigating whether Tai-Chi Chuan contains training components in proprioception. For that, they collected video filming from six Tai-Chi Chuan masters. Their results indicated that the continuous shifting of the center of gravity and a wide range of motion of joints might facilitate the improvement in balance and proprioception.

Wu et al. (2004) addressed the question of how Tai-Chi Chuan can improve balance and posture control by comparing Tai-Chi Chuan gait with regular gait. They study quantitatively characterize the spatial, temporal, and neuromuscular activation patterns of ten healthy young performing normal and Tai-Chi Chuan gaits. The kinematics of the gait was measured using a marker based motion analysis system and two biomechanical force plates. Their results showed that the Tai-Chi Chuan gait exhibited: longer cycle duration (11.9 ± 2.4 vs. 1.3 ± 0.2 s) and a longer duration of single-leg stance time (1.8 ± 0.6 vs. 0.4 ± 0.05 s); larger joint motion in ankle dorsi/plantar flexion (40 ± 9o vs. 20 ± 8o), knee flexion (82 ± 8o vs. 53 ± 10o), hip flexion (81 ± 7o vs. 24 ± 4o) and hip abduction (20 ± 8o vs.0 ± 3o); and a larger lateral body shift (>25% vs.5% body height). The authors claim that the large range of motion found for the Tai-Chi gait maybe be effective in improving both active and passive joint range of motion so that regular practice of Tai-Chi may help improve joint range of motion in the lower extremity. Additionally because they found a mean single-leg stance time for the Tai-Chi gait longer than the entire stance phase of most daily activities, such as walking, going up and down the stairs, the authors speculated that people who practice Tai-Chi Chuan regularly might be able to stand on one leg longer, which help explain the results obtained by Hong et al. (2000).

Wu and Hitt (2005) compared the Tai-Chi gait with slow walking by quantifying the biomechanical characteristics of foot–ground contact during gait. They measured ground reaction force profiles, center of pressure and plantar pressure patterns of the Tai-Chi and slow walk gait of ten healthy young subjects with little Tai-Chi experience. The authors found that the two type of gait were similar in terms of the peak planter pressure and contact area, and peak vertical force.

Mao et al. (2006) also attempted to describe and quantify the plantar pressure distribution characteristics during Tai-Chi exercise and explain the possible beneficial effects of Tai-Chi on balance control and muscle strength when compared with normal walking. In this study however, differently than in Wu and Hitt (2005), they investigated the gait of experienced Tai-Chi practitioners (n = 16; experience of practicing, 8.1 ± 5.7 years). Five typical movements were selected for analysis: the brush knee and twist steps, step back to repulse monkey, wave hand in cloud, kick heel to right, and grasping the bird's tail, representing stepping forward, backward, sideways, up-down, and fixing movements, respectively. They used the Pedar-X insole system to collect the plantar forces during performance of the movements; each insole has 99 sensors and the sampled rate was set at 50 Hz. They measured pressure-time integral, ground reaction force, and displacement of center of pressure. They found that during Tai-Chi movements, the loading of the first metatarsal head and the great toe was significantly greater than in other regions. They also found that the amplitudes of the ground reaction force during the Tai-Chi movements were lower than in normal walking. Compared with normal walking, the locations of the center of pressure in the Tai-Chi movements were significantly more medial and posterior at initial contact and were significantly more medial and anterior at the end of contact with the ground. The displacements of the COP were significantly wider in the medio-lateral direction in the forward, backward, and sideways Tai-Chi movements. The displacement was significantly larger in the antero-posterior direction in the forward movement. According to the authors, these plantar pressure characteristics of Tai-Chi movements may beneficial to intensify the plantar cutaneous tactile sensory input from the first metatarsal head and great toe areas, increase the muscle strength of the lower extremities, and subsequently improve balance control.

Gorgy et al. (2007) used a different approach to investigate the mechanisms of Tai-Chi Chuan and its role in improving balance and posture control. The purpose of their study was to determine the effects of Tai-Chi Chuan as well as two other similar Chinese martial arts (Pa-Koua, Hsing-Hi Chuan) on postural reaction control after an unexpected perturbation. They recruited four groups of six healthy participants: a junior sport group who practiced different sports (soccer, athletics, swimming, basketball, Judo) with a mean frequency of three times per week; a senior sport group who also practiced different sports (soccer, swimming, Judo, cycling) on average twice a week; a martial arts group who had at least five year of experience; and a sedentary group.

Participants were requested to stand in a heel-to-toe position with their eyes either open or closed and were subjected to an unexpected lateral platform translation at two translation amplitudes. Peak displacement of the center of pressure and of the center of mass, and the onset latency of muscular activity of several muscles (tibialis anterior, gastrocnemius, lumbodorsal muscular group, and rectus abdominal) were measured.

Compared with the sport and non-sport participants, the martial arts group showed lower maximal center of pressure and center of mass peak displacements in both the lateral and antero-posterior directions, but no difference was found in the onset of muscular responses.

The authors concluded that the martial arts group used the ankle joint more frequently than the sport and non-sport participants, especially in the eyes-closed conditions and that the better balance recovery in the martial arts group was a consequence of better control of biomechanical properties of the lower limbs (e.g. through muscular response by co-contraction), not a change in the neuromuscular temporal pattern.

4.2.2. Falls

As much as martial arts practitioners train to throw opponents to the ground, they must train to fall to the ground without getting hurt. For that, martial artists train specific fall techniques. For almost a decade, it has been suggested that that martial arts fall techniques can reduce hip impact forces (Sabick et al., 1999). Groen et al. (2007) quantified the role of hand impact, impact velocity, and trunk orientation in the reduction of hip impact force in martial arts techniques. In their study, six experienced judokas performed sideways falls from kneeling height using three fall techniques: using the arm to break the fall, martial art rolling technique with use of the arm to break the fall, and martial art technique without use of the arm. Their results showed that the martial art technique can reduced the impact force by approximately 30%. Impact velocity was also significantly reduced in the falls done with the rolling technique. They found no significant differences between the rolling techniques using and without using the hands. The fact martial arts fall technique can contribute to the reduction of impact forces without using the arm to break the fall provided support for the incorporation of these techniques in fall prevention programs for elderly.

Weerdesteyn et al. (2008) took one step further and investigated whether hip impact forces and velocities in martial arts falls would be smaller than simply using the arm to break the fall in ten young adults without any prior experience after a thirty minutes training session in sideways martial arts fall techniques. They found that martial arts falls had significantly smaller hip impact forces (17%) and velocities (7%). They concluded that training in martial arts techniques is very promising with respect to their use in interventions to prevent fall injuries.

The previous results were supported by a computer simulation study done by Lo and Ashton-Miller (2008). They developed a three-dimensional, 11-segment, forward dynamic biomechanical model to investigate whether segment movement strategies prior to impact can affect the impact forces resulting from a lateral fall. Four different pre-impact movement strategies, with and without using the ipsilateral arm to break the fall, were implemented using paired actuators representing the agonist and antagonist muscles acting about each joint. The results demonstrate that, compared with falling laterally as a rigid body, an arrest strategy that combines flexion of the lower extremities, ground contact with the side of the lower leg along with an axial rotation to progressively present the posterolateral aspects of the thigh, pelvis and then torso, can reduce the peak hip impact force by up to 56%. However, they alerted that a 300 ms delay in implementing the movement strategy inevitably caused hip impact forces consistent with fracture unless the arm was used to break the fall prior to the hip impact. This means that fall techniques needs to be executed very well in order to be efficient in preventing possible injuries and thus need to be constantly trained in soft surfaces.

4.2.3. Wrestling Mats and Shoes

Newton et al. (2002) study is a great example of how biomechanical studies can help preventing injuries during martial arts competition. The purpose of their study was to investigate the effects of perspiration on friction and determine whether there was a significant difference in the coefficient of friction between three different Wrestling shoes (a worn older version shoes that had straight and curved tread on the outer sole; a new and a worn newer version shoes that had circular treads with a circular concave and cup-shaped outer sole) and two different mats (older and newer version of the same type of foam Wrestling mat). They measure coefficient of friction by dragging a weighted shoe over a Wrestling mat surface and measuring the vertical and horizontal forces produced. To simulate a condition when the mat maybe wet by sweat during a competition, the authors smeared a saline solution over the surface of the mat. They found a significant effect of shoe, mat, and wet/dry conditions. In addition, they found significant interactions of shoe by mat, shoe by dry/wet, and mat by dry/wet were observed. The highest coefficient of friction found was 1.54, determined using the worn, new design shoe with the new, dry mat. The lowest co-efficient of friction of 0.60 was found with the old shoe design and the old dry mat. Overall, their results showed that the coefficient of friction was 36% higher for the new Wrestling mat compared to the old Wrestling mat. Application of the saline solution reduced coefficient of friction by 14% compared to the dry condition. Comparison of the mean coefficient of friction for all three shoe types revealed the coefficient of friction for the older design shoe was 23% to 28% lower than the brand new shoe and the worn newer design, respectively. The authors concluded that the coefficient of friction found for the newer mat/new shoe combination had the potential to increase the risk of knee and ankle injuries by fixing the foot more securely to the ground.

Chapter 5

Muscle Activation, Repeatability of Movement, and Reaction Time

Successful performance in martial arts and combative sports requires efficient execution of motor tasks and a high level of perceptual ability. This chapter of the book reviews articles that have investigated martial arts and combative sports using biomechanical tools to quantify muscle activation, repeatability of movement, and reaction time.

5.1. Electromyography

5.1.1. Fast Movements

To date, few studies have focused on the neuromuscular activity of hand strikes. Neto et al. (2007a) compared the electromyographic activity of the triceps brachii, biceps brachii and brachioradialis muscles during palm strikes with and without impacts. Electromyography analyses were done in the time and wavelet domains. Morlet wavelet power spectra were obtained and an original method was used to quantify statistically significant regions on the power spectra. The results both in the time and frequency domains indicated higher triceps brachii and brachioradialis muscle activity for the strikes with impacts. No significant difference was found for the biceps brachii in the two different scenarios. Also investigating the differences in muscle activation between strikes with and without impact, Neto and Marzullo (2009) collected EMG data of the deltoid anterior, triceps brachii and brachioradialis muscles from eight Kung Fu practitioners during both types of strikes. They found that the median frequency of the EMG obtained using wavelet transforms was lower for the strike with impact. The authors suggested that there is a better synchronization of motor units for the strikes with impact performed by the experienced Kung Fu practitioners.

Neto et al. (2008a) reported a kinematical and electromyographic analysis of Kung Fu Yau-Man Palm strikes without impact. An empirical model applied to data obtained by a high-speed camera (1000 Hz) described the kinematical characteristics of the movement. Similar to Neto et al. (2007a), the authors analyzed the electromyographic patterns of the biceps brachii, brachioradialis and triceps brachii muscles during the strike in the time and frequency domains. The electromyography results showed a well developed muscle coordination of the practitioners in agreement with kinematical results. In an attempt to further understand the difference between highly trained martial arts athletes and untrained subjects, Neto et al. (2007b) evaluated the coordination between agonists and antagonists muscles of the arm in a movement of Kung Fu performed by trained and untrained subjects. The authors demonstrated, using wavelet transforms to analyze the data in the frequency domain through time, that the untrained subjects presented much higher undesired co-contractions during the strikes.

McGill et al. (2010) addressed in their study the paradox of muscle contraction to optimize speed and strike force. When muscle contracts, it increases both force and stiffness. Force leads to fast movements, but the corresponding stiffness slows the change of muscle shape and joint velocity. Considering that, the purpose of their study was to investigate how five elite Mixed Martial Arts athletes were able to create high strike force very quickly. They recorded EMG and 3-dimensional spine motion. Participants performed a variety of strikes common in MMA. The results showed that many of the strikes demonstrated what they called a "double peak" of muscle activity. An initial peak was timed with the initiation of motion presumably to enhance stiffness and stability through the body before motion. This appeared to create an inertial mass in the large "core" for limb muscles to "pry" against to initiate limb motion. Then, some muscles underwent a relaxation phase as speed of limb motion increased. A second peak was observed upon contact with the opponent (heavy bag). These results confirmed speculations reviewed previously (3.4.1 Strikes) about how martial artists can generate high values of effective mass to improve their striking power.

Electromyographic studies of kicks are also very rare. Sorensen et al. (1996) examined whether proximal segment deceleration during kicks is performed actively by antagonist muscles or is a passive consequence of distal segment movement, and whether distal segment acceleration is enhanced by proximal segment deceleration. Their results were based on data collected from seventeen skilled Taekwondo practitioners were recorded using a high-speed camera while performing a high front kick, and electromyography recordings from five major lower extremity muscles. Their results indicated that thigh deceleration was caused by motion-dependent moments arising from lower leg motion and not by active deceleration.

Aggeloussis et al. (2007) conducted a research to study the repeatability of electromyography waveforms of major lower limb muscles during the axe kick in Taekwondo. They recorded data from the rectus femoris, biceps femoris, gastrocnemius lateralis and tibialis anterior during 10 successive kicks performed to a fixed target by each participant. The electromyographic activity during the kicks presented coefficient of variation greater than 80%. The authors concluded that only ensemble averages of electromyography waveforms obtained from more than ten kicks may be considered representatives of the muscle function in this type of kicks.

5.1.2. Slow Movements

Tai-Chi Chuan is a martial art known by its slow movement forms. The slow movements of Tai-Chi maybe one of the reasons it helps develop balance and lower body strength. Chan et al. (2003) investigated the muscle activity of a Tai-Chi master asked to perform a sequence a slow motion Tai-Chi push movement. Surface EMG was recorded from the lumbar erector spinae, rectus femoris, medial hamstrings and medial head of gastrocnemius. The authors found low activity of the medial hamstrings and medial head of gastrocnemius and high activity of the lumbar erector spinae and rectus femoris during the push movement. Additionally, they found that both concentric and eccentric contractions occurred in muscles of the lower limbs, with eccentric contraction occurring mainly in the anti-gravity muscles such as the rectus femoris and the medial head of gastrocnemius. The authors suggest that the eccentric muscle contraction of the lower limbs in the push movement of Tai-Chi may help to strengthen the muscles.

Xu et al. (2003) collected EMG of the rectus femoris, semitendinosus, gastrocnemius and anterior tibialis muscles from six Tai-Chi Chuan masters performing another typical Tai-Chi Chuan movement (brush knees and twist steps). Their results indicated that continuous alteration of muscle loading and types of contraction produced different levels of muscular activity, which was helpful to develop muscle strength and endurance. The slow and smooth action of the movement also required well-controlled muscle coordination. All of these effective training factors for neuromuscular control made Tai-Chi Chuan exercise produce particular benefits on postural control.

As we reviewed in (4.2.1 Balance and Postural control) Wu et al. (2004) addressed the question of how Tai-Chi Chuan can improve balance and posture control by comparing Tai-Chi Chuan gait with regular gait. In their study, ten healthy young subjects were tested and surface EMG was recorded from six muscles: tibialis anterior, soleus, peroneus longus, rectus femoris, semitendinosus, and tensor fasciae latae.

They found that the Tai-Chi gait had significantly higher involvement of ankle dorsiflexors, knee extensors/hip flexors and hip abductors, as indicated by significantly higher peak and root-mean-square values of their EMG, longer proportions of action, longer proportions of isometric and eccentric actions and longer proportions of co-activations.

This study showed that ankle dorsiflexors, knee extensors, and hip abductors are activated at a higher level and over a longer duration during Tai-Chi gait than during normal gait. The authors conclude that because of the cumulative effect of the continuous practice, one would expect that Tai-Chi Chuan may be an effective strengthening and endurance exercise for these muscles.

In a similar study Tsen et al. (2007) investigated the vastus lateralis, vastus medialis, bicep femoris and gastrocnemius EMG activity of 11 subjects (five females and six males) during the stance phase of normal walking and a Tai-Chi step. Raw EMG was processed by root-mean-square (RMS) technique using a time constant of 50 ms, and normalized to maximum of voluntary contraction for each muscle. They calculated peak normalized EMG amplitude and co-contraction index. They found that peak EMG and co-contraction values were significantly greater in Tai-Chi stepping compared to normal walking only for quadriceps muscles. The EMG findings of this study showed an increase in levels of knee extensor muscles used during Tai-Chi and may help to support that Tai-Chi exercise helps maintaining good muscle strength of lower limbs by long term practice.

Wu (2008) also compared normal and Tai-Chi gaits but in his study he examined difference between older and younger adults performing these gaits. In his study, EMG recording were collected from 6 young and 6 older subjects (mostly woman). The primary age-related differences found during Tai-Chi gait were during single stance, with elders having significantly shorter activation time in the tibialis anterior (-13%), soleus (-39%), and tensor fascia lata (-21%), activation magnitude in the tibialis anterior (-39%), and co-activation time of the tibialis anterior and soleus (-47%). Compared with normative gait, elders during Tai-Chi gait had significantly higher magnitude (200%-400%) of the tibialis anterior, rectus femoris, and tensor fascia lata muscle activities, and longer duration of co-activation of most leg muscle pairs (130%-380%). The authors also report that older practitioners practice Tai-Chi in higher posture. Finally, they concluded that Tai-Chi gait poses significantly higher challenges to elder's balance and muscular system than does their normative gait.

It seems that many of the benefits of Tai-Chi Chuan come from the fact that the movements are performed in very slow speeds. In order to better understand this, Wu and Ren (2009) investigated the effect of Tai-Chi Chuan exercise performed at different speed on leg muscle activity characteristics in both younger and older Tai-Chi Chuan practitioners. They collected surface EMG of six leg muscles as well as kinematics of lower extremity joints while younger and older subjects practice one Tai-Chi movement (Part wild horse's mane) at fast, normal, and slow speeds. They found that the activation duration of all six leg muscles was significantly longer at slower speed than at faster speed.

Additionally, the durations of isometric, concentric and eccentric actions were either longer at the slower speed or did not change with speed for all six leg muscles. The action of knee extensor was primarily isometric at slower speed, and increased significantly to concentric and eccentric at faster speed. The activation magnitude of posterior leg muscles increased with speed. In general, the old subjects had significantly shorter activation duration and lower activation magnitude in several leg muscles than the young, but were affected similarly by the different speeds.

The authors concluded that the prolonged isometric action of the knee extensors during slower Tai-Chi practice may improve their steadiness in force production and, in turn, contribute to the improvement of balance and prevention of falls. In other words, the general concept that Tai-Chi should be practiced as slow as possible without losing its flow may be effective in improving muscular control and postural balance.

Their findings also suggested that the postural height during Tai-Chi movement does have an impact on the duration and magnitude of leg muscle activation. The authors conclude that it is important to emphasis on the speed rather than the postural height of the Tai-Chi movement for older people.

5.2. Repeatability of Movement

The quantitative analysis of the repeatability of martial arts strikes and combative sports can help the better understanding of practitioners' motor control, as well as serve as a valuable tool for training. Ravier and Millot (1999) analyzed links between decision-making and motor control through Karate techniques. In their study six Karate practitioners had to perform two movements previously trained in three experimental situations. The first was a simple situation in which the subject had no opponent and had to perform the parries (only one possibility). The second was a situation with choice, in which the subject still had no opponent and had to perform one of the parries at the trainer's request (two possibilities). In the third situation the subjects had to choose between the two movements to lock the attack of an opponent. They measured angular variations of elbow and knee joints as well as their duration were measured, using goniometry and showed a significant decrease in the precision of technical skills in the third situation compared with the simple situation.

Sforza et al. (2000) quantified the repeatability of the displacement of thirteen body landmarks of seven Karate practitioners while performing two different basic Karate hand strikes. Subjects were recorded with an optoelectronic computerized instrument (100 Hz) while performing 10 repetitions of straight punches and lunge punches. From the data of each subject, they calculated the average time of execution and the standard deviations of each of the three spatial coordinates x, y, z were computed for each landmark. The results showed that for all subjects, the execution of lunge punch took longer. For both punches and almost all landmarks, the largest repeatability (smallest standard deviation) was found in the vertical direction, while the smallest was found in the anteroposterior direction (direction of movement). Considering all Karate practitioners, they showed that the lunge punch had a total standard deviation about 3 to 6 times larger than that measured during the performance of straight punches. In a similar study, Sforza et al. (2002) studied the repeatability of the Karate front kick when performed by 13 black belt Karate practitioners. The results showed that two experienced athletes presented the lower variability. The best repeatability was found in the horizontal plane, and in general for the hips and head movements.

Roosen and Pain (2007) examined the execution of a popular kicking combination during training in two level of intensity. 3D movement data were captured at 250Hz using a VICON motion analysis system from five Taekwondo practitioners. They found that execution in two different intensity levels cause changes in angle data. Their data suggested that the first kick in the sequence showed the most differences in general. For all kicks, most differences are seen in the central segments. Angles related to the head and to the pelvis show most variability.

5.3. REACTION TIME

In sport science, two types of perceptual abilities have been considered relevant to player's successful performance. One is primitive and constitutes basic sensory functions which are not specific to particular types of sport expertise. The other type of perceptual processing is sport-specific perceptual skills (Mori, 2002). Research on the perceptual abilities of martial artists and combative sports athletes is scarce. Layton (1993), citing one of his previous studies (Layton, 1991), reported that reaction time for hitting a punch bag in response to a sound stimulus were faster for Karate black belts, although the reaction times of the advanced athletes did not differ in proportion to their grades. Kim and Petrakis (1998) the administered Identical Pictures Test, a time-constrained multiple-item test of perceptual judgments, to 50 male and 45 female volunteer Karate practitioners who were classified by skill and belt-rank into three groups. Their results showed that black belts and the women had faster visual-perceptual speed.

Lee et al., (1999) reported that the reaction time to perform a ballistic finger extension movement in was significantly shorter in Kendo, a Japanese martial art whose main focus is sword fighting, (143 +/- 12 ms) and Karate (146 +/- 11 ms) athletes compared to sedentary subjects (176 +/- 12 ms). They suggested that the reaction time is shortened through motor learning in the Kendo1 and Karate athletes who trained for explosive movements. Williams and Elliott (1999) have used more realistic stimuli and tasks to examine expertise anticipation of Karate athletes, which is one important aspect of perceptual skill. In their study, the stimuli were dynamic film displays of Karate athletes performing offensive attacks against the viewer. The film was presented on a large screen to give a real-size view of the performing athletes. The participant's task was a choice reaction time task, where the participants had to respond differently, as soon and accurately as possible, to the attacking positions. In this scenario, anticipating the attacking position from an early part of the video sequence would lead to faster reaction times and/or higher accuracy. The results showed that expert Karate athletes did not react faster, but they were more accurate than the novices. In a similar study Mori et al.

(2002) investigated simple reaction times, choice reaction times and anticipation of Karate athletes. Their results showed significant differences between the Karate athletes and the novices in the choice reaction time task, but not in the simple reaction time.

Pérez (2003) presented an extensive study on reaction time of Karate athletes. In his study, 169 adults practitioners of Karate, and 32 non-practitioners performed a choice reaction time test created using a computer software called SuperLab Pro version 2.0 (Cedrus Corporation, USA). Reaction time task consisted on pushing a button of a computer keyboard as quickly as possible, when a black square appeared on the screen. The black square could be in four different positions, and the subject had to press a different key according with the position. Each subject performed the test one hundred times and the authors calculated their efficiency in the task considering the reaction time and the number of mistakes made. They also collected personal and sport data using a questionnaire, tapping test in 10, 20 and 30 seconds data and hand grip strength data. Comparisons were made considering the subjects' competition specialty (fighting or forms), sport rank (regional, national without medal, national with medal and international) and sex. No significant differences were found in any of the comparisons. They also found that Karate athletes were no different than the general population in the choice reaction time task. They found that reaction time was correlated to tapping and hand grip strength. They also found differences in reaction time depending on gender, concluding that men were faster than women.

O'Donovan et al. (2006) examined simple reaction times, choice reaction times and movement times of thirteen Taekwondo and Kung Fu practitioners and a control group. The experiment consisted of the subjects releasing an initiation button and depressing a stop button 25 cm away. Results indicated that during the simple and choice reaction time tasks the martial artists were no quicker in lifting their hand off a button in response to a sound stimulus, but were significantly faster in moving to press another button. Fontani et al. (2006) examined 18 Karate practitioners, 9 of high experience and 9 of low experience and showed that experience Karate practitioners reacted faster than those of low experience on the simple reaction time test (204 vs. 237 ms).

Reaction time is a fundamental variable for the success of a counter-strike; however, a successful counter-strike also demands precision and power. The studies cited above estimate in different ways reaction time values of martial artists without taking into consideration the functionality of striking. Thus, Neto et al. (2009) proposed a study to investigate reaction time of martial artists during a strike where precision and strike force was also measured. The goal of the study was to compare values of force, precision, and reaction time of several martial arts punches and palm strikes performed by advanced and intermediate Kung Fu practitioners, both men and women. Reaction time values were obtained using two high-speed cameras that recorded each strike at 2500 Hz. Force of impact was measured by a load cell. Precision of the strikes was determined by a high-speed pressure sensor. They found significant negative correlations between the values of force and precision and the values of force and reaction time. Perhaps, martial artists should consider which of these three characteristics is more important for different specific situations, since achieving the best values at the same time in these three variables may be difficult. On the other hand, it can be hypothesized that achieving high values in these three variables may be a consequence of specific training. The results for the comparison between the reaction time data obtained by men and women are in agreement with Pérez (2003), as they showed that men presented faster reaction times than women (women: 342 ± 50ms; men: 243 ± 67ms). Women in the study also presented, on average, lower values of force but higher values of precision than men. Due to the increased number of female martial artists, it is important to further investigate the reasons why women had on average better precision than men and lower values of force and reaction time. The study also demonstrated no significant differences in simple reaction time and precision between the advanced and intermediate participants. Additionally, advanced-level participants presented a higher mean force than intermediate participants. These results suggest that simple reaction time and precision may not be as modified as force with the continuation of martial arts training from intermediate to advanced levels.

As it can be seen, there is controversy in the literature about simple reaction time from martial artists of different expertise. While Layton (1993) found no significant difference in simple reaction time between experienced Karate practitioners of different grades, Fontani, et al. (2006) did find 3rd and 4th dan black belt Karate practitioners to react faster than 1st and 2nd dan black belt Karate practitioners. On the other hand, while Layton (1993) concluded that Karate black belts presented smaller simple reaction times than novices, Williams and Elliott (1999), Mori, et al. (2002) and Neto et al (2009) found no significant difference in simple reaction time between experienced and novice martial arts practitioners. Clearly more research should be done to better understand the possible effects of martial arts training on reaction time.

Chapter 6

FINAL REMARK

The pioneer studies on the biomechanics of martial arts were published in the nineteen sixties and seventies. After these articles were published, several other biomechanical studies have been conducted about martial arts and other related combat sports using a variety of different measures and methods, especially in the last decade. In general, these studies were concerned with: quantifying performance and investigating how to improve it; understanding injury mechanisms and prevention; and investigating potential benefits from training martial arts and combative sports to the general population. This book presented a comprehensive review on this subject. It covered the basics of biomechanics, a brief history of the most popular styles of martial arts and combative sports and a thorough review of over 100 articles and books about the performance of specific hand strikes, kicks, throws and fall techniques, postural control benefits caused by martial arts training and biomechanical investigations of injury mechanisms and prevention.

Although in the last decade there has been an increase in the number of scientific publications about the biomechanics of martial arts and combative sports, there are still many unanswered questions regarding this subject. As we have seen in this book, biomechanical studies of martial arts and combative sports are important not only for athletes and martial artists but also for the general population. At last, I hope that this book helps and inspires scientists, and scientists to be, to do more research in this field.

REFERENCES

Aggeloussis, N., Vassilis, G., Sertsou, M., Giannakou, E. and Mavromatis, G. (2007). Repeatability of electromyographic waveforms during the Naeryo Chagi in Taekwondo. *Journal of Sports Science and Medicine*, 6(CSSI-2), 6-9.

Arlott, J. (1976). *The Oxford companion to sports and games*. Edited by John Arlott. London: Oxford University Press, Oxford.

Atha, J., Yeadon, M. R., Sandover. J., Parsons, K. C. (1985) The damage punch. *British Medical Journal*, 291, 1756-1757.

Baker, J. F., Devitt, B. M., Moran, R. (2010) Anterior cruciate ligament rupture secondary to a 'heel hook': a dangerous martial arts technique. *Knee Surg. Sports Traumatol. Arthrosc,* 18, 115–116.

Benda, B. J., Riley, P. O. and Krebs, D. E. (1994) Biomechanical Relationship Between Center of Gravity and Center of Pressure During Standing. *Ieee Transactions on Rehabilitation Engineering*, 2, 3-10.

Bledsoe, G. H., Hsu, E. B., Grabowski, J. G., Brill, J. D. and Li, G. (2006) Incidence of injury in professional Mixed Martial Arts Competitions. *Journal of Sports Science and Medicine*, CSSI, 136-142.

Blum, H. (1977). Physics and the art of kicking and punching. *American Journal of Physics*, 45, 61-64.

Boey, L.W. and Xie, W. (2002). *Experimental investigation of turning kick performance of Singapore national Taekwondo players*. Proceeding of ISBS, Caceres – Spain, 302-305.

Bolander, R.P., Neto, O.P., Bir, C.A. (2009) The effects of height and distance on the force production and acceleration in martial arts strikes. *Journal of Sports Science and Medicine*, 8(CSSI 3), 47-52.

boxing. (2008). *In Encyclopaedia Britannica*, 2008/12/12, Available from: Encyclopædia Britannica Online: *http://www.britannica.com/EBchecked /topic/76377/boxing*.

Brimacombe, J. M., Wilson, D. R., Hodgson, A. J., Ho, K. C. T. and Anglin, C. (2009) Effect of calibration method on Tekscan sensor accuracy, *Journal of Biomechanical Engineering*, 131, 034503–034504.

Carpenter, C. S. (2005) Biomecânica. Rio de Janeiro: Sprint. In portuguese.

Chan, S. P., Luk, T. C. and Hong, Y. (2003) Kinematic and electromyographic analysis of the push movement in tai chi. *Br. J. Sports Med*, 37, 339–344.

Chiu, D. (2005). *Wrestling: rules, tips, strategy and safety*. (1st edition) New York : Rosen Central. The Rosen Publishing Group Inc.

Chow, D. and Spangler, R. (1982). Kung Fu, History, *Philosophy and Technique*. North Hollywood, CA: Unique Publications.

Conkel, B. S., Braucht, J., Wilson, W., Pieter, W., Taaffe, D., and Fleck, S.J. (1988). Isokinetic torque, kick velocity and force in Taekwondo. *Medicine and Science in Sports and Exercise*, 20(2), S5.

Despeux, C. (1981). Tai Chi Chuan: Arte Marcial Técnica de Longa Vida (5th edition). São Paulo, SP: Círculo do Livro S.A. In Portuguese.

Enoka, R. M. (2008) *Neuromechanics of Human Movement* (4th edition). Human Kinetics Publishers, Leeds.

Falco, C., Alvarez, O., Castillo, I., Estevan, I., Martos, J., Mugarra, F. and Iradi, A. (2009) Influence of the distance in a roundhouse kick's execution time and impact force in Taekwondo. *Journal of Biomechanics*, 42(3), 242-248.

Feld, M. S., McNair, R. E., and Wilk, S. R. (1979) The physics of Karate. *Scientific American*, 240(4), 150-158.

Fontani, G., Lodi, L., Felici, A., Migliorini, S. and Corradeschi, F. (2006) Attention in athletes of high and low experience engaged in different open skill sports. *Perceptual and Motor Skills*. 102(3), 791-805.

Gary, K. (2004) Electromyographic Kinesiology. In Robertson, DGE et al. *Research Methods in Biomechanics*. Champaign, IL: Human Kinetics Publ.

Gorgy, O., Vercher, J.-L., Coyle, T., and Buloup, F. (2007). Coordination of upper and lower body during balance recovery following a support translation. *Perceptual and Motor Skills*, 105, 715 – 732.

Gracie, R (2008). Carlos Gracie: o criador de uma dinastia. Sao Paulo, SP, Editora Record.

Groen, B.E., Weerdesteyn, V. and Duysensa, J. (2007) Martial arts fall techniques decrease the impact forces at the hip during sideways falling. *Journal of Biomechanics*, 40, 458–462.

Group, D. (1977). *Enjoying Combat Sports.* (1st edition) New York, NY, Paddington Press.

Gulledge, J.K. and Dapena, J. (2008). A comparison of the reverse and power punches in oriental martial arts more options. *Journal of Sports Sciences*, 26(2), 189-196.

Halabchi, F., Ziaee, V. and Lotfian, S. (2007) Injury profile in women Shotokan Karate Championships in Iran (2004-2005) *Journal of Sports Science and Medicine* 6(CSSI-2), 52-57.

Hall, S. J. (1993) Biomecânica Básica. Guanabara Koogan, Rio de Janeiro. In Portuguese.

Harter, R.A. and Bates, B.T. (1985). *Kinematic and temporal characteristics of selected Judo hip throws.* Proceedings of ISBS, Del Mar, CA, 141-150.

Higaonna, M. (1985). *Traditional Karatedo-1 Fundamental Techniques.* Tokyo: Minato Research and Publishing Co. Ltd.

Hong, Y., Li, J. X. and Robinson, P. D. (2000) Balance control, flexibility, and cardiorespiratory fitness among older Tai Chi praticioners. *British Journal of Sports Medicine*, 34, 29-34.

Hong, Y. and Li, J. X. (2007) Biomechanics of Tai Chi: A review. *Sports Biomechanics*, 6(3), 453–464.

Huston, R. L. (2008) *Principles of Biomechanics.* Taylor and Francis Group, LLC.

Imamura. R. and Johnson, B. (2003) A kinematic analysis of a Judo leg sweep: major outer leg reap-osoto-gari. *Sports Biomechanics*, 2(2), 191-201.

Imamura, R. T., Hreljac, A., Escamilla, R. F. and Edwards, W. B. (2006). A three-dimensional analysis of the center of mass for three different Judo throwing techniques. *Journal of Sports Science and Medicine*, 5 (CSSI), 122 – 131.

Imamura, R. T., Iteya, M., Hreljac, A. and Escamilla, R. F. (2007) A kinematic comparison of the Judo throw Harai-goshi during competitive and non-competitive conditions. *Journal of Sports Science and Medicine*, 6(CSSI-2), 15-22.

Joch, W., Fritche, P., and Krause, I. (1981). *Biomechanical analysis of boxing.* In K. Morecki, K. Fidelius, K. Kdzior and A. Wit (Eds.), Biomechanics VII-A (343-349). Baltimore, MD: University Park Press.

Kim, H.S. and Petrakis, E. (1998). Visuoperceptual speed of Karate practitioners at three levels of skill. *Perceptual and Motor Skills*,.87(1), 96-98.

Kochhar, T., Back, D. L., Mann, B., Skinner, J. (2005) Risk of cervical injuries in Mixed Martial Arts. *Br. J. Sports Med*, 39, 444-447.

Kutner, N. G., Barnhart, H., Wolf, S. L., McNeely, E., Xu, T. S. (1997) Self-report benefits of Tai Chi practice by older adults. *J. Gerontol*, 52, 242–246.

Knudson, D and Morrison, C. (2002) *Qualitative Analysis of Human Movement*. (2nd edition) Human Kinetics Publishers, Leeds.

Layton, C. (1991). How fast are the punches and kicks of traditional Shotokan Karateka? *Traditional Karate*, 4, 29–31.

Layton, C. (1993) Reaction time + movement-time and sidedness in Shotokan Karate students. *Perceptual and Motor Skills*, 76, 765-766.

Lee, J.B., Matsumoto, T., Othman, T., Yamauchi, M., Taimura, A., Kaneda, E., Ohwatari, N. and Kosaka, M. (1999). Coactivation of the flexor muscles as a synergist with the extensors during ballistic finger extension movement in trained kendo and Karate athletes. *International Journal of Sports Medicine* 20, 7-11.

Li, F., Harmer, P., Fisher, K J. and MCauley, E. (2004) Tai Chi: Improving Functional Balance and Predicting Subsequent Falls in Older Persons. *Medicine and Science in Sports and Exercise*, 36(12), 2046-2052.

Lo, J. and Ashton-Miller, J. A. (2008) Effect of Pre-Impact Movement Strategies on the Impact Forces Resulting From a Lateral Fall. *J. Biomech.*, 41(9), 1969–1977.

Mao, D. W., Li, J. X. and Hong,Y. (2006) Plantar Pressure Distribution During Tai Chi Exercise. *Arch. Phys. Med. Rehabil*, 87, 814-820.

McInnis, B. C. and Webb, G. R. (1971) Mechanics *Dynamics: The Motion of Solids*. New Jersey: Prentice Hall Inc.

McGill, S. M., Chaimberg, J. D., Frost, D. M., and Fenwick, C. M. J. (2010) Evidence of a double peak in muscle activation to enhance speed and force: an example with elite Mixed Martial Arts fighters. *J. Strength Cond. Res*, 24(2), 348-357.

Melone, Jr., C. P., Polatsch, D. B., Beldner, S. (2009) Disabling Hand Injuries in Boxing: Boxer's Knuckle and Traumatic Carpal Boss. *Clin. Sports Med*, 28, 609–621.

Merletti, R. and Parker, P. A. (2004) *Electromyography Physiology, Engineering, and noninvasive Applications*. IEEE Press Editorial Board.

Mori, S., Ohtani, Y. and Imanaka, K. (2002). Reaction times and anticipatory skills of Karate athletes. *Human Movement Science*, 21, 213-230.

Neto, O.P., Bolander, R., Pacheco, M.T.T., Bir, C. (2009) Force, reaction time and precision of Kung Fu strikes. *Perceptual and Motor Skills*, 109, 295-303.

Neto, O. P., Magini, M., and Pacheco, M. T. T. (2007a) Electromyographic study of a sequence of Yau-Man Kung Fu palm strikes with and without impact. *Journal of Sports Science and Medicine*, 6, 23-27.

Neto, O.P., Magini, M., Marzullo, A. C. M. and Pacheco, M.T.T. (2007b). Estudo Eletromiográfico da coordenação entre músculos agonistas e antagonistas do braço durante um golpe de Kung Fu Yau-Man. Terapia Manual, 4, 303-306. (In Portuguese)

Neto, O. P., Magini, M., and Saba, M. M. F. (2007c) The role of effective mass and hand speed in the performance of Kung Fu athletes compared to non-practitioners. *Journal of Applied Biomechanics*, 23, 139-148.

Neto, O. P., and Magini, M. (2008a) Electromyography and kinematic characteristics of Kung Fu Yau-Man palm strike. *Journal of Electromyography and Kinesiology*, 18, 1047-1052.

Neto, O.P., Magini, M., Saba, M. M. F. and Pacheco, M.T.T. (2008b) Comparison of Force, Power, and Striking Efficiency for a Kung Fu Strike Performed by Novice and Experienced Practitioners: Preliminary Analysis. *Perceptual and Motor Skills*, 106, 188-196.

Neto, O. P. and Marzullo, A. C. M. (2009) Wavelet transform analysis of electromyography Kung Fu strikes data. *Journal of Sports Science and Medicine*, 8(CSSI 3), 25-28.

Neto, O. P., Bolander, R., Pacheco, M. T. T., Bir, C. A. (2009) Force, reaction time, and precision of Kung Fu strikes. *Perceptual and Motor Skills*, 109, 295-303.

Ngai, K. M., Levy, F., Hsu, E. B. (2008) Injury trends in sanctioned Mixed Martial Arts competition: a 5-year review from 2002 to 2007. *Br. J. Sports Med*, 42, 686-689.

Newton, R., Doan, B., Meese, M.,Conroy, B., Black, K., Sebstianelli, W and Kramer, W. (2002) *Wrestling. Sports Biomechanics*, 1(2), 157-166.

O'Donovan, O., Cheung, J., Catley, M., McGregor, A. H. and Strutton, P. H. (2006). An investigation of leg and trunk strength and reaction times of hard-style martial arts practitioners. *Journal of Sports Science and Medicine*, CSSI, 5-12.

Oler, M., Tomson, W., Pepe, H., Yoon, D., Branoff, R. and Branch, J. (1991) Morbidity and mortality in the martial arts: a warning. *The Journal of Trauma*, 31, 251–253.

Olsen, P. D., Hopkins, W. G. (2003) The Effect of Attempted Ballistic Training on the Force and Speed of Movements. *The Journal of Strength and Conditioning Research*, 17(2),291-298.

O'Sullivan, D., Chung, C., Lee, K., Kim, E., Kang, S., Kim, T. and Shin, I. (2009) Measurement and comparison of Taekwondo and Yongmudo turning kick impact force for two target heights. *Journal of Sports Science and Medicine*, 8(CSSI III), 13-16.

Pedzich, W., Mastalerz, A. and Urbanik, C. (2006). *The comparison of the dynamics of selected leg strokes in Taekwondo WTF. ACTA of Bioengineering and Biomechanics*, 8, 63-81.

Pérez, O.M.Q. (2003). *El tiempo de reacción visual en el Karate*. PhD. thesis - Universidad Politécnica de Madrid. (In Spanish).

Perry, P. (1982) Sports medicine in China: a group philosophy of fitness. *Physician and Sportsmedicine*, 10(1), 177–178.

Pierce, J. D. Jr., Reinbold, K. A. Lyngard, B. C., Goldman, R. J. and Pastore, C.M. (2006) Direct Measurement of Punch Force During Six Professional Boxing Matches, *Journal of Quantitative Analysis in Sports*, 2(2), Article 3.

Pieter, F., and Pieter, W. (1995). Speed and force in selected Taekwondo techniques. *Biology of Sport*, 12(4), 257-266.

Pieter, W. and Lufting, R. (1994) Injuries at the 1991 Taekwondo world championships. *Journal of Sports Traumatology and Related Research*, 16, 49–57.

Pucsok, J.M., Nelson, K., Ng ED. (2001) A kinetic and kinematic analysis of the Harai-goshi Judo technique. ACTA *Physiologica Hungarica*, 88 (3-4), 271-280.

Ravier, G. and Millot, J. L. (1999). Etude des ajustements moteurs mis en oeuvredans des situations d'apprentissage (kihon) et appliqués en Karate. Science and Sports, 14, 130-136 (in French).

Reid, H., and Croucher, M. (1983). *The way of the warrior: the paradox of the martial arts*. (2nd edition) London: Century Pub.

Roberts, J.B. and Skutt, A.G. (2002). *The Boxing Register* (3rd edition) Ithaca, NY McBooks Press.

Roosen, A. and Pain, M. T. G. (2007). Kinematic changes in the reproduction of a Taekwondo kicking combination. *Journal of Biomechanics*, 40, 455-455.

Sabick, M.B., Hay, J.G., Goel, V.K. and Banks, S.A. (1999). Active responses decrease impact forces at the hip and shoulder in falls to the side. *Journal of Biomechanics* 32, 993–998.

Schwartz, M.L., Hudson, A.R., Fernie, G.R., Hayashi, K. and Coleclough, A. A. (1986) Biomechanical study of full-contact Karate contrasted with boxing. *Journal of Neurosurgery,* 64(2), 248-252.

Serina, E.R. and Lieu, D.K. (1992) Thoracic injury potential of basic competition Taekwondo kicks. *Journal of Biomechanics*, 25(10), 1247-8.

Sforza, C., Turci, M., Grassi, G., Fragnito, N., Pizzini, G. and Ferrario, V.F. (2000) The repeatability of choku-tsuki and oi-tsuki in traditional Shotokan Karate: a morphological three-dimensional analysis. *Perceptual and Motor Skills*, 90(3 Pt 1), 947-60.

Sforza, C., Turci, M., Grassi, G.P., Shiray, Y.F., Pizzini, G. and Ferrario, V.F. (2002) Repeteability of Mae-Geri_Keage in tradiotional Karate: A three-dimensional analysis with black-belt Karateka. *Perceptual and Motor Skills*, 95, 433-444.

Smalheiser, M. (1984). Tai Chi Chuan in China today. Tai Chi Chuan: *Perspectives of the Way and Its Movement*, 1, 3–5.

Smith, M. S., Dyson, R. J., Hale, T. and Janaway, L. (2000) Development of a boxing dynamometer and its punch force discrimination efficacy. *Journal of Sports Sciences*, 18, 445-450.

Smith, P.K. and Hamill, J. (1986). The effect of punching glove type and skill level on momentum transfer. *Journal of Human Movement Studies* 112, 153-161.

Sorensen, H., Zacho, M., Simonsen, E., Dyhre-Poulsen, P. and Klausen, K. (1996) Dynamics of the martial arts high front kick. *Journal of Sports Sciences*, 14, 483-495.

Stojsih, S., Boitano, M., Wilhelm, M., and Bir, C.A. (2008) A prospective study of punch biomechanics and cognitive function for amateur boxers. *British Journal of Sports Medicine* doi:10.1136/bjsm.2008.052845

Swaim, L. (1999). Fu Zhongwen: Mastering Yang Style Taijiquan. North Atlantic, Berkeley, CA.

Taekwondo. (2008). In *Microsoft Encarta Online Encyclopedia*, 2008/12/12, Available from: *http://encarta.msn.com/encyclopedia_761585873/ Tae_Kwon_Do.html*

Takahashi, R. (1992). The application of biomechanics to Judo technique "OKURI-ASHI-BARAI" (Sweeping Ankle Throw). *Sports coach*, 15, 30-33.

Tezuka, M., Funk, S., Purcell, M. and Adrian, M. (1983). Kinetic Analysis of Judo technique. In: *Biomechanics*, VIII-B. Eds: Matsui, H. and Kobayashi, K. Champaign, IL: Human Kinetics. 869-875.

Thompson, C. (2008) *Black Belt Karate.* New Holland Publishers, London, UK.

Thornton, E. W., Sykes, K. S., Tang, W. K. (2004) Health benefits of Tai Chi exercise: improved balance and blood pressure in middle-aged women. *Health Promotion International*, 19 (1), 33-38.

Tsai, Y. J., Huang, C. F., Gu, G. H, (2007) The kinematic Analysis of Spin-Whip Kick of Taekwondo in Elite Athletes. Program and Abstracts of the XXI Congress, International Society of Biomechanics, *Journal of Biomechanics*, 40(2), S615.

Tsang, W. W., Wong, V. S., Fu, S. N. and Hui-Chan, C. W. (2004a) Tai Chi improves standing balance control under reduced or conflicting sensory conditions. *Archives of Physical Medicine and Rehabilitation*, 85(1), 129-137.

Tsang, W. W. N. and Hui-Chan, C. W. Y. (2004b) Effect of 4- and 8-wk Intensive Tai Chi Training on Balance Control in the Elderly. *Med Sci Sports Exerc.*, 36(4), 648-57.

Tseng, S., Liu, W., Finley, M. and McQuade, K. (2007) Muscle activation profiles about the knee during Tai-Chi stepping movement compared to the normal gait step. *Journal of Electromyography and Kinesiology*, 17, 372–380.

Voigt, M. (1989). A telescoping effect of the human hand and forearm during high energy impacts t. *Journal of Biomechanics*, 22(10), 1065.

Vos, J. A. and Binkhorst, R. A. (1966). Velocity and force of some Karate arm-movements. *Nature*, 211, 89-90.

Walilko, T. J., Viano, D.C. and Bir, C.A. (2005). Biomechanics of the head for Olympic boxer punches to the face. *British Journal of Sports Medicine*, 39, 710-719.

Walker, J. D. (1975). Karate Strikes. *American Journal of Physics*, 43, 845-849.

Walter, Donald (December 8, 2003). "Mixed Martial Arts: Ultimate Sport, or Ultimately Illegal?". Grapple Arts. (*http://www.grapplearts.com/Mixed-Martial-Arts-1.htm*).

Weerdesteyn, V., Groen, B. E., Van Swigchem, R. and Duysens, J. (2008) Martial arts fall techniques reduce hip impact forces in naïve subjects after a brief period of training. *Journal of Electromyography and Kinesiology*, 18, 235–242.

Wilk, S.R., McNair, R. E. and Feld, M. S. (1983). The Physics of Karate. *American Journal of Physics*, 51, 783-790.

Williams, A. M. and Elliott, D. (1999). Anxiety, expertise, and visual search strategy in Karate. *Journal of Sport and Exercise Psychology*, 21, 362–375.

Winter, D. A.(1990) *Biomechanics and Motor Control of Human Movement* (2nd edition) Wiley, New York.

Wu, G., Liu, W., Hitt, J. and Millon, D. (2004) Spatial, temporal and muscle action patterns of Tai Chi gait. *Journal of Electromyography and Kinesiology*, 14, 343–354.

Wu, G. and Hitt, J. (2005) *Ground contact characteristics of Tai Chi gait Gait and Posture*, 22, 32–39.

Wu, G. (2008) Age-Related Differences in Tai Chi Gait Kinematics and Leg Muscle Electromyography: A Pilot Study. *Arch. Phys. Med. Rehabil*, 89, 351-357.

Wu, G. and Ren, X. (2009) Speed effect of selected Tai Chi Chuan movement on leg muscle activity in young and old practitioners. *Clinical Biomechanics*, 24, 415–421.

Xu, D., Li, J. and Hong, Y. (2003) Tai Chi Movement and Proprioceptive Training: A Kinematics and EMG Analysis. *Research in Sports Medicine*, 11, 129-143.

Yang, Y. (2005) Taijiquan: the art of nurturing, *the science of power*. Zhenwu Publications, Champaign, Illinois.

INDEX

D

E

F

G

H

I

Q

R

S

T

U

V

W

Y